JOHN ROGERS

TUDOR MILITARY ENGINEER

Oxford University Press, Ely House, London W. 1

GLASGOW NEW YORK TORONTO MELBOURNE WELLINGTON
CAPE TOWN SALISBURY IBADAN NAIROBI LUSAKA ADDIS ABABA
BOMBAY CALCUTTA MADRAS KARACHI LAHORE DACCA
KUALA LUMPUR HONG KONG TOKYO

JOHN ROGERS

TUDOR MILITARY ENGINEER

L. R. SHELBY

ASSOCIATE PROFESSOR OF HISTORY
SOUTHERN ILLINOIS UNIVERSITY

OXFORD
AT THE CLARENDON PRESS
1967

PRINTED IN GREAT BRITAIN

TO
NANCY, REBECCA
AND DEBORAH

PREFACE

WHILE searching for material in the British Museum for a book on medieval English master masons I became interested in the Cottonian collection of fifteenth- and sixteenth-century architectural drawings. Particularly intriguing were several plans of Hull Manor that appeared to have been drafted by the same hand. Curiosity impelled me to look for the man who had produced them; he turned out to have been a sixteenth-century master mason and surveyor of works named John Rogers. Further study of the Cottonian collection disclosed several more drawings which seemed as if they might have been drafted by him. But there were serious problems in securing these attributions, since the drawings presented a number of interlocking puzzles involving drafting and handwriting techniques, dates, and historical circumstances in which the drawings were made. Patient sifting of the evidence has, it seems to me, produced solutions to most of these puzzles, but the reader of this book will, of course, decide for himself how well they have been resolved. While the drawings were grudgingly yielding up their secrets, pursuit of Rogers's career led me to realize that he was a more significant figure in Tudor building than I first thought, and that he was directly involved in important developments in English military engineering. These conclusions have provided the book with a larger framework for piecing together the many small problems connected with Rogers's drawings and with his engineering projects.

In researching and writing this book I have been aided by friends too numerous to mention *in toto*; to those who shall remain unnamed I should like to express my appreciation for their patience in listening to my constant tergiversations regarding this or that puzzle to be found in Rogers's drawings or his career. The assistance rendered by other friends and scholars has been, however, too explicit to go unmentioned. I am particularly grateful to Mr. Howard Colvin and Mr. Martin Biddle, both of whom are writing for volume iii of *The History of the King's Works* chapters that are closely related to portions of the present book. Not only were they willing to read and criticize my book in manuscript, they very kindly shared with me the results of their own research, knowing all the while that my study might well appear before their work could be published. For answering inquiries or for assistance in procuring out-of-the-way materials I am indebted to Dr. D. F. Findlay of Hull, Dr. Marcus Merriman of the University of Lancaster, Dr. A. H. W. Robinson of the University of Leicester, Mr. T. A. Corfe of the Department of Prints, British Museum, and Professor Donald Kelley of the State University of New York. My colleague, Professor Robert Griffin, read the manuscript and suggested a number of stylistic changes which have been incorporated. It goes without saying that any errors of fact, interpretation, or presentation which remain are to be charged against me rather than against those who have so generously allowed me to utilize their special knowledge or services.

Several institutions played a material part in the research for this book. I wish to thank the librarians and other officials for services rendered in the Southern Illinois University and University of Illinois libraries, in the Public Record Office and Kent Archives Office, and especially in the Manuscript Reading Room, the Department of Prints, and the Photographic Service of the British Museum. Appreciation is also due to the American Philosophical Society, the American Council of Learned Societies, and the Southern Illinois University Research Office for grants which made possible the research in these libraries and record offices.

Finally, I wish to express my deepest appreciation to my wife and daughters, the three persons to whom this book is dedicated, for providing a happy and secure family life while at the same time granting that privacy and tolerating that devotion to his work that are the *sine qua non* of the scholar's life.

L. R. SHELBY

Carbondale, Illinois
August 1966

CONTENTS

LIST OF FIGURES

LIST OF ABBREVIATIONS

B.M.	British Museum.
B.S.A.A.B.	*Bulletin de la société académique de l'arrondissement de Boulogne.*
C.P.R.	*Calendar of Patent Rolls.*
Cal. Carew MSS.	*Calendar of the Carew Manuscripts.*
Cal. S.P. Dom.	*Calendar of State Papers, Domestic Series.*
Cal. S.P. For.	*Calendar of State Papers, Foreign Series.*
Cal. S.P. Ireland	*Calendar of State Papers Relating to Ireland.*
Dasent, *A.P.C.*	Dasent, J. R., ed. *Acts of the Privy Council.*
D.N.B.	*Dictionary of National Biography.*
L. & P.	*Letters and Papers, Foreign and Domestic, of the Reign of Henry VIII.* (These are cited by volume, part, and number, rather than by volume, part, and page.)
M.S.A.A.B.	*Mémoires de la société académique de l'arrondissement de Boulogne.*
P.R.O.	Public Record Office
E. 36	Exchequer: Treasury of the Receipt, Miscellaneous Books.
E. 101	Exchequer: King's Remembrancer, Accounts Various.
S.P. 1	State Papers, Henry VIII. General Series.
S.P. 10.	State Papers Domestic, Edward VI.
S.P. 68	State Papers Foreign, Edward VI.

I

INTRODUCTION

THE career of John Rogers, sixteenth-century master mason and military engineer, has been neglected by students of Tudor building and engineering. Whereas Stefan von Haschenperg, a contemporary German engineer in the service of Henry VIII, has been the subject of an illuminating monograph,[1] and Rogers's friend and fellow engineer, Sir Richard Lee, has merited a place in the *Dictionary of National Biography*, Rogers himself has been accorded only three brief biographical notices, along with scattered references in works dealing with sixteenth-century technics.[2] Yet Rogers must rank as one of the foremost English engineers of the time, second perhaps only to Richard Lee. The part played by Rogers in the royal fortification projects at Guines, Hull, and Boulogne would alone justify a more satisfactory biography than he has hitherto received, but there are other reasons for studying the life and work of this man. His career came at an important turning-point in the history of English building, when medieval styles of civil, ecclesiastical, and military architecture were being transformed or set aside under the impact of the Italian renaissance of classical forms. With these changes in style came changes in the status and functions of the personnel responsible for the technical side of medieval building. Although these changes were most dramatic in Italy itself, they were also reflected in England in the careers of men like Rogers and Lee. On this account it will be profitable to investigate Rogers's career as an example of a 'transition figure', whose life and work illustrate the beginning of that transformation of the medieval master masons into the architects and engineers of modern building. In England, to be sure, the development of architecture as a profession separate from

[1] B. H. St. J. O'Neil, 'Stefan von Haschenperg, an Engineer to King Henry VIII, and His Work', *Archaeologia*, xci (1945), 137–55.

[2] John Harvey, *English Mediaeval Architects* (London, 1954), 228, carries the story of Rogers's career only to 1545. A. H. W. Robinson, *Marine Cartography in Britain: A History of the Sea Chart to 1855* (Leicester, 1962), 18–19, 147–8, provides a more complete account of Rogers's life, but unfortunately continues that life into the 1570s, when in fact Rogers died in 1558. Whitworth Porter, *History of the Corps of Royal Engineers* (London, 1889), i. 24–29, completely garbles the facts of Rogers's career.

the building crafts was a long, slow process that was not formally completed until the nineteenth century. Yet a study of the status, functions, and technical skills of men like Rogers suggests that the sixteenth century witnessed an important turning-point in this long story. Not only was Rogers a part of this 'professionalization' of certain elements in the medieval building crafts, but his career also illustrates that specialization of functions which ultimately separated the profession of architecture from that of engineering. Again one must be cautious not to overemphasize the distinction between architecture and engineering in sixteenth-century England; many if not most building technicians still practised both, as had their medieval predecessors. But a few men like Rogers and Lee were developing into specialists who were widely known for their skill and knowledge in engineering, and particularly in military engineering.

The fact that in his career Rogers had little to do with ecclesiastical or civil architecture is not to say that he was *ab initio* uninterested in this kind of building. It is more likely that the career opportunities which opened up to him directed his interests into military architectural and engineering works. These opportunities lay almost entirely within the royal service, and, in the case of Rogers, developed within the particular office of the surveyor of the works. The history of this office can be traced back into the fourteenth century, where it was associated primarily with administrative and clerical aspects of royal building.[1] By the time of John Rogers, however, the office was being filled with men who had received technical training in the art of building, and who were expected to assume technical supervision of construction as part of their duties.[2] Thus the office was being transformed into a new kind of position which involved both administrative and technical capacities. Furthermore, since it was often associated with military projects, it offered particular opportunities to those building technicians in the royal service who were devoting themselves especially to military architecture and engineering. As a result, some of the English surveyors of works were emerging as equivalents of those continental engineers who were currently being employed for their specialized knowledge and skills in military fortification.[3] Since Rogers was one of the most

[1] H. M. Colvin, ed., *The History of the King's Works*, (London, 1963), i. 168–70.

[2] This development may have begun around 1532 with the appointment of the king's Master Carpenter, James Nedeham, as Surveyor of the King's Works. Harvey, op. cit. 190; cf. John Summerson, *Architecture in Britain, 1530 to 1830*, 4th ed. (Harmondsworth, 1963), 5.

[3] Odet de Selve, French ambassador to England during the years 1546–9, recognized

prominent of the surveyors in his own day, his career provides a useful case study of the duties involved in that office at a time when it was undergoing important modifications.[1]

If we speak of Rogers as a specialist because his career as a surveyor of works involved him almost entirely with military fortifications, we must not overlook the many different roles he played and functions he performed while acquiring his reputation as a military engineer. In various stages of his career he served the king as a working mason, master mason, land surveyor, architectural and topographical draughtsman, architect, military engineer, hydraulic engineer, technical building consultant, administrative clerk, military spy, and diplomatic attaché. Thus another aspect of Rogers's career impresses itself upon us—the high esteem in which Henry VIII held Rogers, as reflected by the multiplicity of duties which the king imposed upon one of his favourite building consultants. At least in matters dealing with military fortifications, Henry came to rely more on Rogers than upon his high-ranking military commanders. He consulted frequently with his surveyor, in person and through the exchange of letters and architectural or topographical drawings. As a result, the study of Rogers's life gives us at least brief glimpses into one of the features of Henry's character that is not normally treated in works dealing with the political, religious, and marital problems of his reign. In his later years Henry displayed a real interest—at times almost a passion—for military architecture. We may be permitted, then, to take occasion in this biography of Rogers to look with interest at Henry, who could not only devise grand schemes for the conquest of France, but who could also concern himself with technical details of how best to fortify a hill-top or a haven.

As noted above, one of the chief means by which Henry and his officials communicated technical building information was through the exchange of 'plats', that is, architectural and topographical drawings. Many of

this when he once described Rogers as 'un ingénieur et maître des œuvres qu'on appelle à Londres surveilleur'. *Correspondance politique de Odet de Selve*, ed. Germain Lefèvre-Pontalis (Paris, 1888), 341.

[1] It will not be our purpose to work out a full-scale definition of this office or to trace its history beyond the confines of Rogers's life. The surveyorship was to have a great future, for out of it was to emerge by the end of Elizabeth's reign the office of Surveyor-General of the Royal Works—the most exalted position in English architecture and building, and one that would be filled in time by such luminaries as Inigo Jones and Christopher Wren. But we must await the history of this office in the third volume of *The History of the King's Works*.

these plats have survived and are now deposited in the Cottonian collection in the British Museum. Of these drawings, a surprisingly large number can be attributed to John Rogers—perhaps more than for any of his contemporaries, and certainly more than can be associated with any earlier English master mason. These drawings, many of which have hitherto remained anonymous, undated, and unpublished, acquire an additional interest because they can now be directly associated with a particular historical figure. Furthermore, they tell us something not only about the career of John Rogers, but also about the royal projects upon which he worked and to which the drawings themselves refer. Since all of these buildings have disappeared, one of the more important by-products of this study will be the reconstruction of the appearance of these royal works, through analysis of the drawings as the chief evidence for such a reconstruction. Finally, the plats hold considerable interest because they illustrate methods of technical drawing as practised by a sixteenth-century English engineer. As such, they form a link in that long development of a satisfactory graphic language for architects, engineers, and cartographers. Since the sixteenth century was one of the turning-points in this process—which lasted from late medieval into recent modern times—we may investigate with special interest these plats by Rogers, in order to see what part he played in the history of technical drawing.

The present study thus has a dual purpose: to illustrate certain developments in sixteenth-century building through the career of an important Tudor military engineer, and to investigate a number of problems connected with the technical drawings produced by that engineer. These two tasks will lead us along several highways and byways that should be of interest to various kinds of readers, including local, military, and even diplomatic historians, as well as those who are particularly interested in the history of engineering and of technical drawing.

In the arrangement of the book there will be an attempt to achieve this dual purpose simultaneously. While Rogers's career unfolds in chronological order, we shall engage at suitable points in analytical discussions of the problems of authorship, dating, and historical background of those drawings which can be attributed to him. Having pursued this chronological/topical scheme to the end of Rogers's life, we shall in the last two chapters evaluate his place as a military engineer and platmaker within the context of sixteenth-century engineering and technical drawing.

II

NEW BULWARKS FOR GUINES

In the late spring of 1541 John Rogers was appointed master mason at Guines Castle, the English stronghold on the southern edge of the Pale of Calais. From this date until his death in 1558 Rogers's career can be traced in a fairly continuous line; yet little is known about him before he became master mason at Guines. This is not untypical of the story of many medieval English building technicians, where evidence for their careers becomes available only after they had achieved a position of prominence within their craft. An additional difficulty with John Rogers is that his name was quite common in the sixteenth century, so that we must beware of confusing the subject of this study with other men of the same name.[1] For instance, it has been noted that 'in 1523 John Roger, freemason, was among the inhabitants of Lavenham, Suffolk, in the Muster Roll for Babergh Hundred, and was assessed at 10*s*.'.[2] Possibly this was the man who later appeared at Guines, though other evidence suggests that Rogers the master mason hailed from London rather than Suffolk.

Twelve massive volumes of works accounts for Hampton Court during Henry VIII's reign contain scattered references to a mason by the name of John Rogers; it seems more than likely that this was the future master mason. Unfortunately, these references do not tell us a great deal about Rogers, but because of the scarcity of evidence regarding his early career we must attempt to extract from them all the information possible. During the accounting period of 19 October to 15 November 1533 Rogers worked for 3*s*. 4*d*. per week as a 'lodgeman', that is, a stonecutter working in the masons' lodge or workshop.[3] He was next to be found working as a stonesetter from 19 December 1534 to 13 March 1535.[4] Beginning with

[1] The most famous John Rogers of the day was, of course, the Protestant theologian and martyr, the first to be executed under Queen Mary in 1555.

[2] John Harvey, *English Mediaeval Architects* (London, 1954), 228, citing the Grimsthorpe MSS.

[3] P.R.O., E. 36/237, p. 255. He worked less than a week during this accounting period, for he was paid only 2*s*. 2*d*.

[4] E. 36/238, pp. 85, 101, 121; E. 36/243, pp. 11, 27, 47.

this latter date, he worked for a week as a lodgeman and then disappeared from the works until 15 May, when he returned as a stonesetter and worked until 10 July.[1] He then absented himself until November, when he was paid 12*s.* for carving special stones 'by convention', that is, under contract.[2] Once again he disappeared from the scene, to return in May of 1537 and work until the middle of August as a lodgeman.[3] After this latter date he worked no more at Hampton Court.

From these scattered facts of Rogers's somewhat chequered career at Hampton Court a number of conclusions can be drawn. There is one definite reference to him as John Rogers of London, freemason.[4] This presumably means that London was his chief place of residence, if not his birthplace, and the presumption is strengthened by the fact that in later years, when it was decided to reward Rogers for his services to Henry VIII, he was given properties in London. Furthermore, the evidence from the records at Hampton Court can be used for making an educated guess about the date of his birth. By 1533 he had clearly finished his apprenticeship as a mason, for he was receiving the full wages of a journeyman when the first references to him are found in the accounts. On the other hand, the fact that some masons received higher wages than he did suggests that he was not yet at the top of his craft and that he was probably still a fairly young man. Since—so far as we know—he first achieved the position of master mason in 1541, we may suppose that he was between thirty and forty-five years of age at that date, which would put his birth-date some time during the period 1495–1510. This range of possible years for his birth correlates satisfactorily with the date of his death in 1558, for there is little reason to suppose that at that time he was much over sixty or much under fifty years of age.

From the summer of 1537 until the spring of 1541 Rogers disappears from the royal building accounts. This does not necessarily mean, of course, that he was entirely dissociated from royal building works, for not all of the accounts of the period have survived. But a search of those that

[1] E. 36/238, p. 293; E. 36/240, pp. 343, 407; E. 36/243, pp. 67, 147, 185.

[2] E. 36/240, p. 57: 'Also payd to John Rogers of London ffremason ffor in taylyng the beccates of the dore in the Kynges great watchyng chambre wyth the beccattes of the chymney in the same chambre, wyth the Kinges badge and the quenys letter by convencyon.' L. F. Salzman, *Building in England down to 1540* (Oxford, 1952), 113, explains that 'beckets' were 'carved bosses at the end of a hood-mold, on either side of a door'.

[3] E. 36/244, pp. 198, 238, 304.

[4] See n. 2 above.

are extant in the Public Record Office has turned up not a single reference to Rogers. This gap in the record of Rogers's career is unfortunate, for it was during these years that he advanced from being an ordinary working mason to become master mason at Guines, and we should very much like to trace exactly the stages in that process. But even if the details are missing, the outline of Rogers's early career is discernible. No doubt he served an apprenticeship in the masons' craft before the middle 1530s, at which time he acquired experience as a working mason in the king's service. This much may be asserted—that Rogers's educational background, and presumably his social background as well, lay in the medieval traditions of the masons' craft. It is important to keep this background in mind as we trace the changes in military engineering and in the status and function of the engineer which are reflected in Rogers's later career. That he became something different from the usual medieval master mason was the result of these changing conditions in English royal building which the study of his career gives us the opportunity to investigate.

The town and castle of Guines, located six miles south of Calais, constituted one of the more important outlying fortresses in the Calais Pale. Guines had been in English hands since its capture from the French in 1352, during the early part of the Hundred Years' War. The place had been fortified since 998, when a Danish invader and conqueror named Sifrid built an earthen mound from which he could dominate the countryside.[1] In the latter quarter of the twelfth century the Comte d'Ardres, Baudouin II, replaced the wooden structure atop the mound with a circular stone donjon.[2] During the next 150 years the French made further additions to the castle, which had achieved a considerable size by the time the English captured it in 1352. This is evidenced by the fact that soon afterwards the English dug 221 rods of moat around the castle in order to strengthen its defensive position.[3] During the fifteenth century they continued to enlarge and fortify the castle, adding both internal buildings and rooms as well as external bulwarks that were furnished with gunports.

[1] Lambert d'Ardres, *Historia Comitum Ghisnensium*, ed. Joh. Heller, *Monumenta Germaniae Historica, Scriptores* (Hanover, 1879), xxiv. 567, no. 8. It is interesting that this earliest portion of Guines Castle should be the last to survive, for all the later additions have disappeared; there remains today only the mound of earth hedged about with modern buildings.

[2] For a brief, rather ambiguous, but contemporary description of this building, which later came to be called 'La Cuve' (The Tub) because of its curious appearance, see ibid. 596, no. 76. Cf. the modern account by Abbé Haigneré in the *Dictionnaire historique et archéologique du Pas-de-Calais. Arrondissement de Boulogne* (Arras, 1882), iii. 23.

[3] P.R.O., E. 101/173/7.

Because of these bulwarks, it has been suggested that 'Guines was one of the earliest fortresses under English control to be adapted to the needs of artillery warfare'.[1] Nevertheless, the castle at Guines—as well as the other fortifications in the Calais Marches—became a source of concern to Henry VIII and his councillors in the late 1530s. During this period considerable tension was mounting between the English and French Governments. Should war erupt between the two nations, the Pale of Calais would be a focal point in the offensive and defensive strategy of both sides. Considering these possibilities, it was disconcerting to English officials to find the various fortifications in the Pale in an alarming state of disrepair. In the spring of 1535 it had been necessary to bring masons from England for 'working in the castle of Guines and taking down the old wall over against Pirton's Bulwark which was ready to fall'.[2] Apparently these precautionary measures were insufficient, for in February of 1536 the chamberward wall fell into the moat, and it was reported that the 'old walls both of Calais and Gysnes are in great danger by reason of the last frost'.[3] Repairs at Guines were begun immediately, so that by July the fallen wall had been restored to a height of 31 feet.[4] This work was carried out under the direction of William Lelegrave, who died in late July or early August. His place as surveyor of the works at Calais was taken by Richard Lee.[5]

For the next several years after his appointment in 1536 Lee was engaged in bringing the fortifications in the Calais Pale up to suitable standards. At first this seems to have been mostly a matter of repairing the older works that had been inherited from the French or that had been constructed by the English since the conquest of Calais in 1347.[6] But apparently it was not until 1539 that serious considerations for new fortification works got under way.[7] With trouble brewing between the English and French, Henry and his ministers felt the need for repairing and augmenting the fortifications at Calais and the outlying English castles. Indeed, the Pale was part of a massive scheme for fortifying England and English territories in preparation for a possible war against France. In 1539 Cromwell sent a remembrance to Parliament setting forth a list of

[1] H. M. Colvin, ed., *The History of the King's Works* (London, 1963), i. 453.

[2] P.R.O., E. 101/540/4. Spelling modernized; henceforth in the text of this book quotations from manuscript sources will be rendered with modern spelling and punctuation.

[3] *L. & P.* x. 395.

[4] Ibid. xi. 144.

[5] Appointment of Lee: ibid. xi. 385 (13).

[6] For a documentary history of the various works in the Calais Pale, 1347–1485, see Colvin, op. cit. i. 423–56.

[7] Cf. the 'Suggestions for further repairs of the fortifications at Guisnes in addition to some already made. . . .' *L. & P.* xiv. i. 901.

twenty-five places in England which were to be fortified, as well as Calais, Hammes, and Guines 'on the other side of the sea'.[1]

The spring of 1540 saw additional works undertaken at Guines, as indicated in the building accounts surviving from this period. During February and March numerous labourers and carters were kept busy gathering materials for making bricks for the new works. Twenty-three bricklayers went to work at the castle in April, and during the next month they were joined by ten masons and six carpenters. Approximately this same number of craftsmen continued to work at Guines throughout the summer of 1540. Some of the carpenters were building a mill, while the masons and bricklayers were engaged in the construction of vaumures, or outer walls, for the keep and for the chamberward of the castle.[2] In October the number of masons decreased to three and the bricklayers to seventeen. By November all of these men had returned to England for the winter. Except for a few carpenters and labourers, the works at Guines remained shut down from November until the end of February. Then during March 1541 thirteen masons and twenty-three bricklayers, along with ten carpenters and a horde of labourers, returned to work at the castle. The chief object of attention during this month was the 'Mill Tower' which was being taken down by the bricklayers, while the masons were hewing hardstones for the foundations of a new tower on this site.[3] Unfortunately, the comptroller's accounts which have provided the foregoing information end on 26 March, and the succeeding accounts of the works at Guines have not survived. It is necessary therefore to turn to other sources in order to determine the nature of the work being executed.

By April 1541 it was evident that the English were engaged in more than the usual annual repairs of the fortifications in the Calais Pale. The French and imperial ambassadors in London, naturally anxious over news about military build-up in this English stronghold, reported to their respective monarchs on two major projects that were being planned at Guines. One was to fortify the town itself; the other was to build four new bulwarks for the castle.[4] In connexion with the first project, Henry was warned by a Portuguese engineer that the town could not be fortified

[1] *L. & P.* xiv. i. 655.

[2] P.R.O., E. 101/206/9, pp. 193 ff.

[3] E. 101/206/13, p. 531.

[4] *L. & P.* xvi. 711 and 733. Technically speaking, the new works proposed for Guines were bastions rather than bulwarks. In this study I shall make no distinction between these terms, because the word bulwark was used in sixteenth-century English documents in reference to what was, in fact, a bastion. It would

effectively because of two nearby hillocks which would overlook the proposed walls. This opinion did not please the king, who dismissed the engineer in disgust, 'calling him an ass who did not understand his business.'[1] Apart from the insight into Henry's opinion of his own technical knowledge and competence which this anecdote provides, the reference to a *Portuguese* engineer provides a clue to the solution of a long-standing puzzle connected with a drawing of the castle and town of Guines.[2] The drawing contains inscriptions in a conglomerate language of French and what has been thought to be Italian or Spanish, but may now be recognized as Portuguese. One of the inscriptions clearly connects the drawing with the anecdote about Henry and the Portuguese engineer, for it states that a bulwark at one of the corners of the town would be useless because of the low ground at that point.[3] It is likely then that the drawing constitutes a graphic survey of the town and castle as they existed in the early spring of 1541, and, possibly, that it was drafted by the Portuguese engineer to illustrate the point that he was trying to make with Henry VIII.

Whoever this Portuguese engineer may have been, he was probably serving at Guines in an advisory capacity only, for Richard Lee, as surveyor of works in the Calais Pale, would have been responsible for the overall planning and execution of the new fortifications at Guines. That Lee took an active interest in the works is shown in 'A Remembrance in what state the king's majesty's works are at this present', which was apparently drawn up by him in April 1541.[4] Here Lee reveals a detailed knowledge of the works and specifically mentions a new wall, 'the foundation of which wall began at my being there'. Lee's interest is understandable, for the works at Guines appear to have been the most prominent in the Calais Pale at this time. Yet there were matters in Calais itself and in the other forts of the Calais Marches which required Lee's attention, and this accounts for the fact that a master mason was now appointed at Guines to direct these new works. There had been a master mason of the

be misleading then for me to preserve the distinction while quoting from documents that do not.

[1] *L. & P.* xvi. 733.

[2] B. M. Cott. MSS., Aug. I. ii. 23; reproduced in *A Commentary of the Services and Charges of William Lord Grey of Wilton*, ed. P. de M. G. Egerton, Publications of the Camden Society, xl (London, 1847), following p. xxiv.

[3] Cf. the transcription of these inscriptions in *L. & P.* xxi. ii. Appendix, 33.

[4] B.M., Royal MS. 7. C. xvi, f. 80. The date of 16 April is specifically mentioned in the document, but the year is not given. The editor of *L. & P.* xv. 527 has placed it in 1540, but internal evidence relating to the works at Guines suggests that it should be dated rather in the spring of 1541.

king's works in Calais itself almost continuously since 1349.[1] William Baker, currently holding that office, had directed repairs at Guines in the summer of 1539,[2] but the working season of 1540 found no master mason directly involved there.[3] With the arrival of spring in 1541 and the formulation of plans for new bulwarks for the castle and ramparts for the town, it was realized that a technical supervisor would be needed in addition to the attention which Richard Lee could give to the new projects. It was thus that Rogers appeared upon the scene as master mason of the king's works at Guines.[4]

We should like to know Rogers's whereabouts just before his arrival at Guines and, especially, whether or not he had previously held the position of master mason in the royal service. Unfortunately, this information has not been forthcoming. One thing, however, is fairly certain: Rogers was not raised from the ranks of the working masons at Calais, Guines, or other forts in the Pale, for his name appears in none of the relevant account books that are fairly complete for the period just preceding Rogers's appointment.[5] One may infer either that he had already held a supervisory post somewhere in the royal works, or that he had come to the attention of royal officials as a promising mason, for he was now entrusted with an important position on one of the major royal fortification projects of that year.

We should also like to know whether Lee and Rogers had known each other previous to Rogers's arrival at Guines, or whether this was the very beginning of that friendship and professional association which periodically reunited them in engineering projects for more than a decade. Lee and Rogers, Rogers and Lee: their names reappeared like a refrain for the next several years, while they steadily rose in prominence, singly and together, until by the end of Henry VIII's reign they were the two most trusted military engineers in the service of the king.

This desire to know more of the immediate antecedents of the association

[1] See Colvin, op. cit. ii. 1054, for a list of the men who held the post from 1349 to 1485.

[2] E. 101/206/6.

[3] E. 101/206/9 lists no such position, nor does any one of the working masons receive higher wages than his fellow men. Probably Richard Lee gave the instructions to the relatively small crew of masons and bricklayers who were executing repairs on the keep and vaumure of the chamberward.

[4] In June Rogers is mentioned simply as the king's master mason (*L. & P.* xvi. 917), but a document dating from August specifically refers to him as the master mason at Guines (ibid. 1108). He was not the master mason of Calais as suggested by Harvey, op. cit. 24 and 228, for that office was still held by William Baker. See *L. & P.* xix. i. 812 (8).

[5] E. 101/206/2, 6, 8, 9, 10, 12, 13, and 14.

of Lee and Rogers arises out of important considerations regarding Rogers's knowledge of military architecture and his skill as an architectural draughtsman. Rogers later constructed fortifications at Hull in Yorkshire that were very similar in design to those which had shortly before been started at Guines. The question arises as to whether the defences at Hull represented a further projection of his own ideas on military architecture, or whether he merely borrowed designs which Richard Lee had developed at Guines. Another problem involving the relationship between the two men centres upon four contemporary architectural drawings (Pls. 1–4) that pertain to the projects under way at Guines during the summer of 1541. All four drawings appear to have been drafted by the same hand, and there is good evidence for believing that either Lee or Rogers was the draughtsman. But which one? The answer in either case has important consequences for an understanding of Rogers's development as an architectural draughtsman.

Since knowledge of the design of the fortifications at Guines is largely dependent upon these drawings, the two problems outlined above are closely related. The best hope of solving them lies in an intensive analysis of the drawings and of the documentary evidence pertaining to the drawings. Even this may not produce categorical assurance regarding authorship of the plans, but at least in the process some things should be learned about the purpose and character of the drawings, as well as about the fortifications to which they referred.

In the spring of 1541 Henry VIII sent the Earl of Southampton and John Lord Russell on an inspection tour of the fortification works in the Calais Marches. On 8 May they informed the king of their arrival at Guines and took the opportunity to give their first impressions of the new works there. At the same time they mentioned certain architectural drawings or plats relating to the fortifications, and this reference may be seen to have important implications for the analysis of the four drawings mentioned above. The letter from the two officials has unfortunately been severely damaged, but with considerable effort (and the aid of an ultraviolet lamp) the following text can be read:

> . . . We arrived here [yesterday after-]noon and incontinent saw your plats w[hich your] mason brought hither, and that done that day we viewed afull your castle as you re[quested.] And this morning and afternoon we have eftsoons viewed the same again, considering t[he] said your plats and them together, and to [be]

plain with your grace, by the faith, duty, [and] allegiance which we bear unto your Majesty, that and you and six of the wisest men within [the] realm had been here a fortnight together, y[ou] could have devised your plats no better th[an they] have done. And as far as our wits [can] compass, we cannot invent or find [. . .] how conveniently to add to or d[etract any-]thing of what they have devised. . . .[1]

It is clear from this account that more than one plat was available for comparison with the works in progress, although there is no positive statement as to who made the drawings. The reference at the beginning of this passage to 'your plats' should not be construed to mean that Henry VIII had drawn them, for it is later remarked that the king himself could have devised no better plats than 'they' have done.[2] One naturally supposes that Southampton and Russell were referring here to the technical building officials at Guines—that is, Lee and Rogers—who would have been conducting the visiting dignitaries on the inspection tour. The letter specifically states that a mason brought the plats to the two officials, and this was very likely the master mason, John Rogers.[3] At other points in the letter Southampton and Russell referred to the surveyor of the works and mentioned that they were going to ask Mr. Lee to make a plat of the marshes near Calais. It is thus at least plausible to associate Lee and Rogers with the plats seen by the two officials, and to suppose that these plats had been produced in planning the new works at Guines. But this still leaves open the question of whether these plats were the four drawings of Guines that are reproduced in Plates 1–4. To answer this question one must compare the drawings with other contemporary evidence to see if they fit into the 'plausible' scheme just presented.

The most informative source in this matter is an anonymous document endorsed 'A writing concerning the castle of Guines and the fortification thereof'.[4] There is no date given, but internal evidence suggests that it was composed in the spring of 1541, when the plans were being laid for reworking the castle and for providing more adequate defences for the town. Although the beginning of the document has been lost, the closing

[1] P.R.O., S.P. 1/165, f. 192.

[2] I have interpolated the word 'they' in the passage where the writing has been effaced, but the verb is plural, and since there is insufficient space at this point on the document for two or more proper nouns, it is reasonable to insert the plural pronoun.

[3] The editor who calendared this document for *L. & P.* xvi. 813 rendered this sentence thus: '. . . saw your plattes w[hich your master] mason brought hither'. Much as we should like to accept this reading, the lack of space for the word 'master' on the illegible portion of the page makes it difficult to do so.

[4] B.M., Royal MS. 7. C. xvi, f. 81.

sentence, 'Good Master Knyvet, recommend me to my good Lady Knyvet', suggests that it was addressed to Edmund Knyvet, sergeant-porter to Henry VIII and holder of various minor royal offices. It can be inferred from the document itself that Knyvet had earlier visited Guines and discussed the new projects with the author. As to the identity of this author, his familiarity with the castle and town, and his authoritative manner of describing the plans for refortification incline one to guess that he was the surveyor of the works, Richard Lee. His intentions in writing the document were to review the discussion which he had held earlier with Knyvet and to provide information on works already under way and on those yet to be started. To accomplish these goals he took Knyvet on an imaginary perambulation of the town walls and castle; we can do no better than to follow him on this tour.

The town of Guines lay immediately south of the castle, and since the author's first concern was to describe the town walls, he began at the south-east corner of the castle and proceeded clockwise around the walls until he had returned to his starting-point. First he mentioned that a new bulwark was being built in front of 'Pirton's Bulwark', which itself constituted the south-east corner of the castle. From there he moved southward to the bulwark 'that shall be made' at the south-east corner of the town. Turning westward, he came to the 'west gate', where another bulwark had been proposed. Further onward was the south-west corner of the town, where a third bulwark on this south wall was to be built. From this point he moved to the north-west corner, where he suggested that a bulwark might be built if the king desired, although it was not absolutely necessary because of the protection afforded by 'Whettle's Bulwark' and other neighbouring bulwarks. Turning eastward he arrived at the bulwark that was currently being built in front of the keep of the castle; from there he returned to the starting point before Pirton's Bulwark. The 'tour' around the works of the castle itself was somewhat briefer. There was mention of the reworking of an older bulwark at the north-west corner, which project corresponded to the 'new tower' that was under construction at the north-east corner. This new tower, the author remarked, adjoined 'the base court by the mill as you come from Calais'.[1]

[1] It will be remembered from the comptroller's accounts (E. 101/206/13) that in March 1541 workmen were taking down the Mill Tower and preparing stone for a new tower at that corner of the castle.

This verbal description can be closely correlated with the four drawings of the castle and town of Guines, but before doing so it will be useful to provide a nomenclature by means of which the drawings can easily be cited. Using a term that was commonly applied to architectural and topographical drawings in the sixteenth century, we shall refer to the plan of the simpler town walls as Plat A (Pl. 1), to the plan of the bastioned town walls as Plat B (Pl. 2), to the plan of the castle as Plat C (Pl. 3), and to the elevation of the castle as Plat D (Pl. 4).

The inscriptions on Plat A enable us to identify the castle keep, as well as Pirton's Bulwark, Whettle's Bulwark, and the east and west gates of the town walls. Since none of the new bulwarks are shown—that is, neither those under construction nor those yet to be constructed—it is clear that this plat was intended to show the state of the defences *before* any of the new works had got under way. This is a rather important point, for it must be realized that the town already possessed a moat and ramparts—however rudimentary in form—before the new works were planned in 1541. Although Plat A shows the moat clearly enough, the precise character of the old town walls does not emerge very clearly from the drawing. In this case the plat executed by the Portuguese engineer is rather more helpful, for that drawing shows the walls consisting simply of earthen ramparts surmounted by wooden palisades.

The English engineers clearly faced a considerable task in transforming these medieval defences into something more adequate for modern warfare. What they proposed to do is set forth in Plat B, which shows six new bulwarks protecting the corners and flanks of the town walls. Although the main lines of the walls remain the same, the moat has been widened and modified to conform to the configurations of the new bulwarks, and the ramparts have been enlarged. Plats A and B thus present 'before and after' plans of the town's defences. That they were intended to be viewed in conjunction with one another is indicated by the fact that they were drawn on exactly the same scale and—as may be concluded from the strong similarity in drafting technique—by the same hand. But the question arises, does Plat B represent work as planned or work as completed? In considering sixteenth-century architectural drawings of buildings that are no longer extant one must be careful to distinguish between those drawings that are surveys of completed structures and those that are proposals for works that may or may not have been executed according to the proposal.

A study of the evidence indicates that Plat B was the latter type of drawing. Comparison of the plat with the 'Writing concerning the castle of Guines' shows a close correlation between the verbal and the graphic descriptions of the new bulwarks. Indeed, the correspondence is so close that the plat might well have been drafted to accompany and to illustrate the 'Writing', except for the fact that on the drawing no distinction was made between those bulwarks under construction and those merely proposed at the time of the 'Writing'. If the drawing was definitely not a graphic survey of work in progress, the problem still remains as to whether or not the works were actually executed as designed. We remember that at the time of the 'Writing' two bulwarks were already under construction—the one before the keep and the one before Pirton's Bulwark. There are several references in contemporary documents to the Three-Cornered Bulwark; it can easily be identified on Plat B with the bulwark to the south-west of the keep. The 'new bulwark before Pirton's Bulwark' can be seen on Plat B to the south-east of the castle. This new bulwark was so closely identified with Pirton's Bulwark that in time it too was called by that name.[1] Contemporary documents indicate that these bulwarks were were completed, while a later (undated) 'Plot of Guins' (Pl. 5) suggests that they were finished in much the same form as shown on Plat B.

Regarding the four other major bulwarks for the town walls as described in the 'Writing' and shown on Plat B, it appears that they were never completed, and indeed probably never begun; at least, documentary evidence is lacking for their actual construction. This argument from silence is supported by 'The Plot of Guins', which shows only a small portion of the town walls, but this portion is in the same form of a simple earthen rampart topped by a wooden palisade that is to be seen on the plan produced by the Portuguese engineer before work on the bulwarks had begun.[2] On the other hand, a view of the siege and capture of Guines by the French in 1558 was printed in Rome in that same year, and this view shows a pentagonal fortification with bulwarks at the five points that bears some resemblance to the plan of the town walls as seen on Plat B.[3]

[1] For references to the Three-Cornered Bulwark and Pirton's Bulwark, see *L. & P.* xvi. 877 and 917; xviii. i. 640 and 833.

[2] For references to the Portuguese drawing, see p. 10, n. 2 above.

[3] The print is entitled 'Ritratto della fortezza di Ghines, presa per forza de englesi dallo essercito del Re Christianissimo'. A copy is in the British Museum (Map Room, C. 7. e. 4. [31]); for other copies, see R. V. Tooley, 'Maps in Italian Atlases of the Sixteenth Century, Being a Comparative List of the Italian Maps Issued

Since 'The Plot of Guins' is undated, it is at least possible that the bulwarks in question were built between the time that this 'Plot' was drawn and the Italian print of Guines was published in 1558. Yet a closer study of the print reveals that it is almost completely untrustworthy in its depiction of the appearance of Guines. The draughtsman has placed the castle just south of the town, when in fact it was just north. He has shown 'Il Castel vecchio' as a square tower keep surrounded by a rampart with round towers at the four corners. Around this square is a moat, and around this moat, 'La Piazza d'intorno al Castel vecchio'. Surrounding the whole of this is the pentagonal rampart with bulwarks at the five points. One of the bulwarks is named 'Il Baluardo della Cuna', another 'Il Baluardo Grande'. The town is depicted as quadrangular, with four massive round towers at the corners and intervening curtains with gates at the mid-points.

From this description it will be seen that the print is an utter misrepresentation of the overall scheme of both the castle and the town; yet it is evident that the view is not entirely a figment of the artist's imagination, but is based on a certain amount of factual information. 'Il Baluardo della Cuna' may be recognized as the name of the (three-cornered) bulwark before the keep, which keep was known to the French as 'La Cuve'.[1] 'Il Baluardo Grande' probably referred to the large new bulwark that had been built near Pirton's Bulwark. Furthermore, the locations of 'La Porta, che va verso Ardres' and 'La Porta, che va verso Hames' are fairly correct indications of two gates of the town. Yet it is hardly conceivable that a draughtsman who possessed the skill displayed in this drawing would have made so many gross errors if he had himself seen the town and castle. He must therefore have based his view on second-hand information, which either was garbled when he got it or which he greatly misunderstood when he tried to present it in graphic form. At any rate, it is clear that the print can hardly be considered trustworthy, independent evidence for the appearance of Guines in 1558, regarding either the detail or the overall scheme of the town and castle. We are then free to maintain our conclusion

by Lafreri, Forlani, Duchetti, Bertelli and Others, Found in Atlases', *Imago Mundi*, iii (1939), 32. This print appears to be the one referred to by Abbé Lefebvre, *Histoire générale et particulière de la ville de Calais, et du Calaisis* (Paris, 1766), ii. 309–10, where he mentions that 'un plan de la ville de Guînes, levé en 1558 par le même Ingénieur Italien, qui a fait celui de Calais, & imprimé aussi à Rome, chez *Duchetti*, représente cette Place, comme une des plus fortes de l'Europe'.

[1] See p. 7, n. 2 above.

that 'The Plot of Guins' is more correct in suggesting that—except for Pirton's Bulwark and the Three-Cornered Bulwark—the bulwarks and ramparts for the town as shown on Plate B were never executed.

Plat C is a plan of Guines Castle which likewise shows designs for the new bulwarks along the south side of the castle. But even a cursory comparison of Plats B and C reveals fundamental differences in both the form and placement of the bulwarks as shown on the two plans. On Plat C a bulwark almost encircles the keep, and there is a small one beyond the moat, whereas neither of these is to be seen on Plat B. On the other hand, Plat C shows no Three-Cornered Bulwark to the south-west of the castle, nor is the design of Pirton's Bulwark at the south-east corner of the castle the same as that shown on Plat B. Since we have already demonstrated the trustworthiness of Plat B in representing these two bulwarks as actually built, one must suppose that Plat C represents designs that were never executed.

This conclusion creates doubts about using Plat C as evidence for the other works which were under way at Guines and which seem to be indicated at various places on the plan by means of gunports similar to those shown on the south-side bulwarks. In particular the north-east corner suggests some elaborate reworking of the older defences. This corresponds with the reference in the 'Writing' to a new tower under construction at that point, 'which is by the mill as you come from Calais'.[1] Other documents refer to the construction of a 'bulwark that responds to the Mill Tower', or sometimes simply Mill Bulwark.[2] But the form of this bulwark as seen on Plat C differs considerably from the later 'Plot of Guins' (Pl. 5). Since by now one is hardly inclined to trust Plat C when there is conflicting evidence, it seems better to follow the later drawing and to suggest that the Mill Bulwark had a trefoil plan similar to that of Pirton's Bulwark.

Yet another problem regarding Plat C may be noted. The imperial ambassador in London, it will be remembered, had reported in the early spring of 1541 that four new bulwarks were being planned for the castle. But in the ensuing months there are references in the documents only to the construction of the Three-Cornered Bulwark, Pirton's Bulwark, and the Mill Bulwark, while a progress report of 7 August 1541 mentions that the 'three bulwarks could not be better placed'.[3] One begins to suspect either that the ambassador had received misinformation, or that the

[1] B.M., Royal MS. 7. C. xvi, f. 81. [2] *L. & P.* xviii, i. 640, 833. [3] Ibid. xvi. 1075.

decision had been made to build three new bulwarks instead of four. This suspicion is confirmed by 'The Plot of Guins', which shows the addition of only the three major bulwarks named above. On the other hand, at least six additional bulwarks can be counted on Plat C, if one assumes that new works are indicated wherever gunports are shown. Besides the four bulwarks along the east and south sides of the castle that have already been discussed, there are the two small ones on the west side—one adjacent to Whettle's Bulwark and the other at the north-west corner of the castle. Since there is no other evidence for special works at these points, it becomes clear once again that Plat C contains designs for new bulwarks that were never executed as shown on the plan.

These numerous discrepancies in Plat C call for special comment. The drawing itself is carefully executed, and it constitutes the best surviving plan of the older parts of Guines Castle. This is not a mere preliminary sketch, but a seriously considered plan for reworking the defences of the castle. Yet it is a plan that was entirely superseded by the designs for the new bulwarks as shown on Plat B. This conclusion has two important implications. It establishes the relative dates of the two drawings, and it shows that in the process of designing the new works at Guines, the engineers changed their minds regarding the proper form that the bulwarks should take. Now it so happens that the form which was finally accepted and which is represented in Plat B was the one that John Rogers later used in his work at Hull. Furthermore, as will be seen, the designs at Guines and Hull merely represented stages in the development of a new type of fortification which later received fuller expression in Rogers's work at Boulogne and in Lee's work at Berwick.

Plat D poses few problems. It is a view of the south elevation of the castle, and it plainly shows the medieval defences before the new works were undertaken. In this respect it corresponds to Plat A of the town walls, and indeed may have been drafted about the same time.

It may be helpful to summarize the conclusions reached about the drawings thus far. Plats A and D constitute surveys of the town walls and castle of Guines as they existed before the refortification of 1541 began, and thus they predate the other drawings in subject-matter and probably in point of time as well. Their purpose was to show the current conditions of the town and castle; as such, they provided bases for comparison with those other drawings that presented schemes for reworking the defences of

Guines. Plat C shows just such a scheme for reconstructing new bulwarks at strategic points around the castle, but in fact the scheme was never carried out, either in the placement or in the form of the bulwarks. Plat B is a proposal for protecting the corners and longer sides of the town walls by means of six large bulwarks. Again, this scheme seems not to have been carried through except for the construction of the two bulwarks nearest the castle. But these two were constructed in their form and placement just as shown on the plat.

If the relative chronology and the purpose of these drawings have been established, and if their trustworthiness as records of the work actually executed has been determined, there still remains the tricky problem of deciding who it was that drafted the plats. Our analysis of the drawings has shown that they can be closely related to the condition of the works at Guines in the spring of 1541, so that they may very well have been the plats seen by Southampton and Russell on 7 May. The evidence of the letter from these two men admits the possibility and even suggests the probability that these plats were produced by either Richard Lee or John Rogers. Lee would be a natural choice as the draughtsman because of his position as chief technical official in the Calais Pale and the man ultimately responsible for the design of the fortifications there. Another consideration may be brought to bear here. If Lee did compose the 'Writing concerning the castle of Guines', then the close correlation between this document and Plat B suggests that he produced the one as well as the other. Since the drawing style of the four plats is so similar, it would be natural to assume that if he produced Plat B, he also drafted the other three plats.

But it is precisely on this matter of style that a serious problem arises, for as we shall see, the drafting technique of these drawings shows strong similarities to architectural drawings later produced by John Rogers. This means that if the plats of Guines were drawn by Richard Lee, then Rogers had either previously acquired a style very close to Lee's before coming to Guines, or else he learned to imitate Lee's techniques while he worked with Lee at Guines. The first alternative is somewhat unlikely in an age when English architectural drawing varied considerably from one hand to the next, because there were as yet no standardized drafting techniques.[1] The second alternative raises the question of whether Rogers would have so

[1] A mere leafing through the large collection of contemporary drawings in the Cottonian collection will substantiate this remark.

completely come under the influence of Lee's drafting style in the relatively short period of his stay at Guines, a period which could have lasted no more than six or seven months. Finally, it should be noted that we are to some extent groping in the dark regarding Lee's drafting style and its possible influence on Rogers. In due course in this study we shall be able to speak fairly confidently of Rogers's style, because of the numerous drawings attributable to him; but we cannot in the same way describe Lee's style, and for the opposite reason—we do not possess the same confidence in attributing to him any very large number of the extant contemporary architectural drawings.[1]

To suppose that it was Rogers and not Lee who drafted the plats in question, one would have to assume a close collaboration between the two men in the production of the drawings, since Lee was the man ultimately responsible for designing the new works at Guines, if not for drafting the plats that would illustrate those designs to the king and his officials. There is some evidence that, indeed, the collaboration between Lee and Rogers was close. On 2 July they dispatched under their joint signatures a letter to Henry VIII, in which they described the progress of the works up to that time.[2] The letter has been badly mutilated, but enough remains to show that Lee and Rogers were both directly involved, not only in the supervision of the workers, but also in the planning of the work. It does not seem unreasonable then to suggest that Rogers, as master mason, was called upon to produce the drawings of Guines, with Lee, as surveyor of the works and Rogers's immediate supervisor, looking over his shoulder—literally or figuratively—while the drawings were being executed.

[1] At least one drawing, Cott. MSS., Aug. I. i. 56, can be definitely ascribed to Lee because of the inscription on it: 'This plott made by Mr. Lee An°: 25° Hen: 8 [1533/34].' But this is a topographical plat of Orwell Haven executed primarily in water-colour; although two forts are shown, the plat does not give an adequate notion of Lee's technique in making architectural drawings with pen and ink. Mr. Marcus Merriman of the University of Lancaster has suggested to me that several plats of Scottish fortresses (Belvoir, Leics., Belvoir MS. Letters, vol. 11, ff. 54–65) may have been drawn by Lee. I would not attempt to pronounce on this possibility at the present moment. In the most recent study on Lee's career, Iain MacIvor ('The Elizabethan Fortifications of Berwick-upon-Tweed', *Antiquaries Journal*, xlv (1965), pls. XXIII and XXIV) publishes two drawings (P.R.O., MP. F/137; and Hatfield House, Cecil Papers, Maps 1/25) which were probably drafted by Lee in 1558 and 1561 for the new works at Berwick that he himself was directing. The whole question of Lee's career as an architectural and topographical draughtsman needs a thorough investigation of the sort that I am attempting to give to John Rogers, in order to ascertain which of the extant drawings may be his and to delineate his characteristics and qualities as a draughtsman.

[2] P.R.O., S.P. 1/166, f. 92 (*L. & P.* xvi. 956).

Admittedly, many of the foregoing arguments are based on circumstantial evidence, and they merely show that Rogers *could* have been the master mason, that he *could* have drawn the plans referred to, and that these plats *could* have been the same plans which are now preserved in the British Museum. A more direct proof of authorship lies in the comparison of the drafting technique of these plats with other drawings which can be shown to have been made by Rogers. But this must wait until the authorship of these later drawings is established.

As master mason at Guines Rogers was no doubt kept busy in the task of directing the actual work of construction; but there were other duties involved in this position besides that of design and technical supervision of building. Henry VIII took a personal interest in the progress of royal building projects, so that much of the correspondence coming from Calais was filled with reports on the work of fortification. In these letters the administrative officials sometimes cited the opinions of Lee and Rogers on technical problems.[1] But Henry also enjoyed hearing the reports first hand from his technical supervisors, as evidenced by the fact that Rogers returned to England three times in the summer of 1541 and rendered account on the progress of the works directly to the king and Council.[2] It is evident from this that Rogers was already gaining favour with the king; his star would continue to rise for the next several years, until he finally reached the position as surveyor of works at Boulogne where he could argue with impunity against his commanding officers, thanks to the high esteem in which the king held his surveyor.

One of the reasons why men such as Rogers could gain the ear of the king was that they could serve him in so many different ways. In those days when specialization in the art and craft of building had not yet completely ruled out the possibility of one man filling the multiple roles of designer, technical supervisor, administrator, and building contractor, even a craftsman such as Rogers, with his valuable 'know-how' in building, could become a trusted adviser of the king. As a military engineer his technical knowledge could be utilized not only in fortifying castles for the king, but in spying out and helping to destroy the castles of the king's enemies. In his career in the royal service Rogers was several times called upon to perform this latter task. During the summer of 1541 the French were busily fortifying places just outside the Calais Pale. In order to

[1] *L. & P.* xvi. 994. [2] Ibid. 917, 992, and 1075.

ascertain their progress in this matter, Rogers was sent in July 'to mark the works at Arde and Mountory' where the fortifications were proceeding under the direction of the Great Master of Flanders, Comte de Roeulx.[1] Proudly Rogers reported that these French works were 'nothing in comparison' with the king's works at Calais and Guines.

Rogers's career at Guines was a brief one. A considerable amount of work remained to be done on the castle;[2] yet Henry pulled Rogers off the project in the autumn of 1541 and sent his master mason northward to supervise the fortifications that were soon to get under way at Hull. It was a step upward in the royal service for Rogers, who relinquished his position as master mason at Guines to become surveyor of the king's works at Hull.

[1] Ibid. 1036. In this reference 'Mountory' (i.e. Montoire) is noted as 'being in distance but one mile from Arde'; a sixteenth-century map B.M., Cott. MSS., Aug. I. ii. 71) shows it on the east side of Ardres.

[2] In early August it was predicted that the fortifications at Guines would be finished by October (*L. & P.* xvi. 1076); nevertheless a complaint was registered in June 1543 that the ditch before the castle, the 'bulwark that responds to the Myll tower', and Pirton's Bulwark would not be finished until Michaelmas, unless more masons were sent. Ibid. xviii. i. 640.

III

HULL AND THE DEFENCE OF THE NORTH

THE town of Kingston upon Hull was chosen in 1539 as one of the twenty-five sites in England to be fortified for war against any of the Catholic powers—papal, imperial, French, or Scottish—which might attempt through arms to bring England back into the Catholic fold. Since a war with France appeared the most likely of these possibilities, greatest attention was given immediately to the defences of the Pale of Calais and the southern coast of England. The fortification of Hull was not undertaken seriously until the autumn of 1541, when the occasion presented itself for Henry personally to inspect the defences of the town and to recommend the works that seemed necessary. For some time Henry had been trying to wean his nephew, James V of Scotland, away from his close connexions with the French throne. By the summer of 1541 Henry had at least succeeded in obtaining a promise from James that he would meet the English king in York, where they might discuss some of the difficulties that separated the two monarchs as well as the two realms. In September Henry kept the rendezvous at York, but James failed to appear. This affront to the king of England was not borne lightly; war with Scotland was now a likelihood, and did, indeed, eventuate in the summer of 1542.

Meanwhile, it was advisable to make ready the defences of the north; to this task Henry and his councillors applied themselves with customary energy. Henry had visited Hull briefly in September on his way northward to meet James. When the Scottish king failed to appear, Henry returned to Hull for a closer inspection of the town. During the first few days of October he and his councillors studied the town walls that had been constructed during the later part of the fourteenth century. In 1321 Edward II had granted to the burgesses of Hull a licence 'to strengthen their town with ditches and a wall of stone and lime and to crenellate the wall'.[1] In spite of this royal privilege, the townspeople required a long time to build their walls, as evidenced by the licence renewals and extensions of time

[1] *C.P.R., 1321–4*, 7.

which fourteenth-century kings granted to the 'mayor, bailiffs, and good men of Kingston-upon-Hull' in 1327, 1348, 1355, 1376, 1382, 1390, and 1396.[1] Apparently the main work of fortification was carried out after the grant of 1382, for in 1384 the townspeople received a royal exemption from sending anyone to represent them in Parliament, 'on account of the fortification of the town, which, the king is informed, they are executing at great expense'.[2] This work of fortification was pushed forward by Michael de la Pole, the wealthy merchant of Hull who had recently become the Duke of Suffolk. Describing the effect of Michael's leadership, the sixteenth-century antiquary, John Leland, remarked: 'And yn his tyme the toune was wonderfully augmentid yn building, and was enclosid with diches, and the waul begon, and yn continuance endid and made al of brike, as most part of the houses of the toun at that tyme was.'[3]

These medieval walls of Hull have entirely disappeared, but a sixteenth-century bird's-eye view of the town (Pl. 6), which apparently is an Elizabethan copy of a much earlier drawing, probably gives a fairly accurate notion of the appearance of the walls around 1400, after they had been completed.[4] In any case, the drawing can be checked and supplemented by the views of Hull which Speed published in 1610 and Hollar in 1640.[5] By 'averaging' this pictorial evidence it is possible to get an idea of what the defences of Hull were like when Henry visited the town in 1541; this in turn will help us to understand the changes which he and his Privy Councillors proposed at the time of his visit. The walls surrounded only three sides of the town, for it had been thought that the east side was sufficiently protected by the Hull River. There was, however, a small brick tower on the east side of the river. In the north and west walls of the town were

[1] Ibid., *1327–30*, 33; *1348–50*, 87; *1354–8*, 155; *1374–7*, 274; *1381–5*, 175; *1388–92*, 362; and *1396–9*, 40.

[2] Ibid., *1381–5*, 175.

[3] *The Itinerary of John Leland*, ed. Lucy T. Smith (London, 1907), i. 48.

[4] Charles Frost, *Notices Relative to the Early History of the Town and Port of Hull* (London, 1827), 80, concluded that the drawing is a copy of a plan which dates from around 1350. The drawing certainly does contain some archaic features—the North Bridge over the Hull River as well as Hull Manor inside the town are not shown, and there are open fields inside the perimeter of the walls—which indicate an early date for the town as shown on the drawing. But the fact that the walls *are* completed suggests a date for the earlier drawing that is closer to the end of the fourteenth century.

[5] John Speed, *The Theatre of Great Britaine* (London, 1611), map following p. 81. Hollar drafted three views of Hull around 1640. One of these is reproduced in Pl. 7. On these and other contemporary drawings relative to Hull, see Thomas Sheppard, *The Evolution of Kingston-upon-Hull, as shewn by its Plans* (Hull, 1911), 44–48.

some seven gates, extending from North Gate by the Hull River to Hessle Gate by the Humber. (Cf. Fig. 1 with Pls. 6 and 7.) On the south wall was the Watergate, which opened on to a 'landing place' on the bank of the Humber. Between these various gates a considerable number of towers, some square and some round, were placed rather closely together.

After Henry and his councillors had studied these defences carefully, they proposed a number of changes in a set of resolutions dated 4 October 1541. 'For the fortification of the haven and town: it is thought meet that a bulwark be made at the Watergate as it pleases the king's majesty to devise the same.'[1] It is clear from this and subsequent decisions relating to Hull that one of the main concerns was to provide more adequate protection for the haven, which was located in the confluence of the Hull and Humber rivers. This first resolution called for the landing place in front of Watergate to be fortified with a bulwark which could protect the entrance to the haven that was close by. This was not an entirely new idea, for the Elizabethan copy of the earlier drawing (Pl. 6) indicates that this landing place had for some time been defended with several cannons. The proposed bulwark would fairly adequately protect the haven entrance, but there was need for some means of protecting the haven itself from the east side of the river. The second resolution read: 'Item, that the little round brick tower on Holderness side be enlarged to bear the chain and to beat the haven, and a guard to be established in it for the sure keeping of the same.' Here again the suggestion was one of improving upon current practice, for the Elizabethan copy shows a chain across the haven, though not attached to 'the little round brick tower', which is shown a bit further upstream. The next resolution called for another fortified point for protecting the haven. 'Item, the brick gate at Northend to be mured up and made platform to beat the flank of the town of the one side and the flank of the haven of the other side, according to the king's device.' This item clearly referred to the North Gate, which was to be walled up and provided with artillery that could be fired westward along the town wall and eastward across the haven. But it is not clear from the wording whether the North Gate itself was to serve as the 'platform' for mounting the guns, or whether an earthen platform was to be built in front of the gate.

Henry and his councillors next turned their attention to the defence of the town proper. 'Item, the corner tower to be made larger out to answer

[1] P.R.O., S.P. 1/167, f. 83. The other resolutions shortly to be quoted are on fols. 83–84ᵛ.

to the brick gate next the haven and to respond to the gate where Constable hangeth.' This tower must have been the round one at the north-west corner of the town, where Hollar's view (Pl. 7) shows it to have been rather insignificant for such an important position. This would account for Henry's desire to make it more imposing, in order to answer to the brick gate next the haven (that is, North Gate) and to respond to the gate where Constable hangs (that is, Beverley Gate).[1] So that this corner of the town's defences might further be strengthened, the resolutions next called for 'the gate where Constable hangeth to have a barbican made to defend the gate and flanks'. While this decision was made to strengthen Beverley Gate, it was concluded that Hull had too many gates to fortify all of them properly. 'Milgate' (apparently meaning Hessle Gate) was to be left open because it provided convenient access to the ferry of Hessle and to pastures belonging to the Hull townspeople, but all other gates than those previously mentioned were to be walled up. This resolution would probably have included Low Gate, Postern Gate, Myton Gate, and the gate at the end of Blanket Row.[2] (See Fig. 1.) Finally, the king's majesty's house (that is, Hull Manor) was 'to be made to serve as a citadel and a special keep of the whole town in such form as it shall please the king's majesty to devise and appoint the same'.

The foregoing measures to strengthen the defences of the town were to be undertaken at the king's expense. At the same time there were certain projects which were to be undertaken by the townspeople themselves.[3] They were to make ramparts 'in all places convenient with the soil of the town'.[4] They were also to cleanse the moat surrounding the town and renovate the sluices which fed water into the moat. Finally, they were to furnish iron pieces of artillery—but no brass pieces—for the defence of the town, as well as portcullises for all gates that were not to be walled up. In

[1] The gruesome reminder which provided the epithet for this latter gate was the body of Sir Robert Constable, whom Henry had caused to be hanged in chains at Hull in June 1537 for his part in an uprising that had occurred during the previous January. *D.N.B.* iv. 969–70.

[2] Of these, only Lowgate and Myton Gate are to be seen on the Elizabethan copy (Pl. 6). Hollar's view (Pl. 7) shows all of these gates, but the bridges at Postern Gate and 'Blanket Row Gate' are missing (although the piles are still there), which indicates that these two gates were no longer in use in 1640. They may have come into being after the walls were originally built, but were 'mured up' as a result of the resolutions in 1541 and remained closed thereafter.

[3] For these, see *L. & P.* xvi. 1232 (2).

[4] This apparently referred to the earthen ramparts which backed up the masonry towers and curtains on the north and west walls. Pl. 6 indicates such ramparts rather awkwardly, but the view by Speed, loc. cit., reveals quite clearly their character as sloping earthworks.

addition to these separate tasks to be performed by town and Crown, there was one mutual undertaking. The townspeople had just petitioned the king for a bridge across the Hull River leading towards Holderness. Henry agreed to build it, provided that the town would contribute money and labour for its construction and would henceforth keep it in repair at the town's own expense.

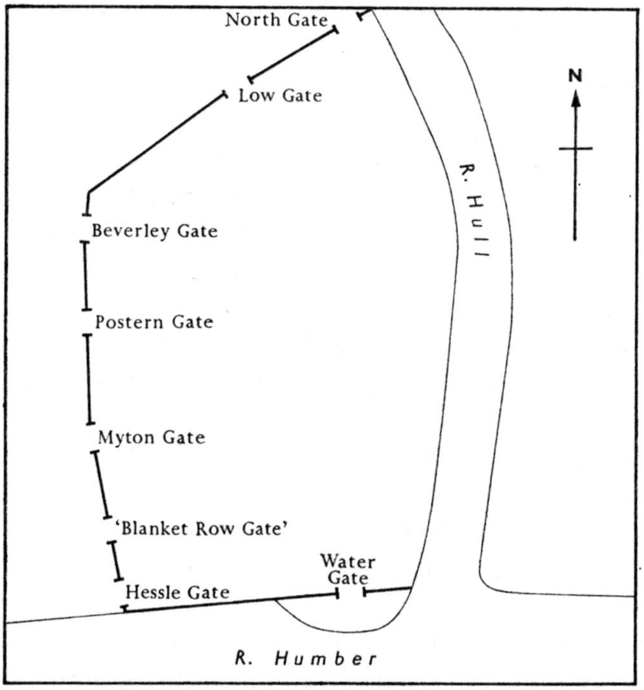

FIG. 1. *Schematic Plan of Town Walls and Gates of Hull c. 1540*

Once the plans had been made in early October 1541 for strengthening the defences of Hull, the necessary military, administrative, and technical machinery was set in motion for carrying through with those plans. By February 1542 a new captain and lieutenant had been appointed to Hull to command the garrison, and John Rogers had been transferred from Guines to serve as surveyor of the works for the new fortifications. Indeed, Rogers was probably drafted for the work at Hull soon after Henry's visit in October, for he is reported to have brought to the king an account of the expenses at Hull before Christmas of 1541.[1]

[1] *L. & P.* xvii. 140.

The administrative arrangement at Hull is interesting because it reveals the interweaving of building technicians with military and administrative officers in the surveillance of the king's works. The captain, Sir Richard Long, and his lieutenant, Michael Stanhope, were primarily responsible for commanding the garrison at Hull, but they were also to oversee the king's works and to have charge of the payroll for both the garrison and the construction workers. Rogers, as surveyor of the works, was the chief technical supervisor of construction, and he was to be assisted in this work by a master mason and a master carpenter. But he was also given administrative duties when he was appointed as comptroller of the works, a post that made him responsible for auditing the account books kept by the paymaster. To assist him in this task he was given a clerk. This point, by the way, will prove to be of considerable interest when we come to the analysis of some drawings attributable to Rogers. The paymaster was required to draw up two account books of particulars after each payroll. Both of these books were to be audited and signed by Rogers, as well as the master mason and master carpenter. One book would then be turned over to Captain Long and the other retained by the paymaster. Furthermore, Rogers and the two master craftsmen, 'if they be not sick', were to sit with the paymaster at each pay.[1] This elaborate check is not surprising in view of the size of the payroll: by 21 April 1543 the works at Hull had cost the king £16,351. 12*s.* 10*d.*[2]

Where had this large sum of money been expended in little more than a year and a half? The resolutions of 4 October 1541 had called for a relatively small-scale reworking of the town's defences, with the main attention devoted to Watergate, North Gate, and Beverley Gate. Unfortunately, the account books of particulars which Rogers as comptroller audited and signed have not survived, and the extant official correspondence relating to the royal works at Hull is meagre, so that there is little documentary evidence for reconstructing either the procedures or the end results of those works. But the views by Speed and Hollar indicate that the proposed works for the above-mentioned gates were not carried through.[3]

[1] Ibid. for all of these administrative details.

[2] Ibid. xviii. i. 433.

[3] Admittedly, the evidence for the 'bulwark at the Watergate' is contradictory. Hollar's bird's-eye view (Pl. 7) shows no bulwark on the landing place before the gate, although his low-angle perspective at the top of this same print does show a low-lying fortification *outside* the walls of the town. Whatever the explanation of this inconsistency, the form of the bulwark as shown by Hollar would probably have been too advanced for English fortifications in the year 1542.

On the other hand, both views show massive fortifications which had been built across the river from Hull but which had not figured in the resolutions of October 1541. Obviously something had caused Henry to change his mind and to undertake a much larger project than he originally intended. That something was probably John Rogers. As we have seen, he was already at work at Hull by Christmas 1541. In an estimate of wages for the king's works at Hull, dated 1 February 1542, it was reported that there were sixty bricklayers who 'shall be set on work upon the bulwark next unto the Humber by the Jetty as fast as the foundation can be dug'.[1] Since this bulwark by the jetty was not the same as the bulwark in front of Watergate, clearly the original plans had been changed by the time this estimate was made.

Although we cannot trace in the documents the events which led to this change of plans, we can be certain of its outcome, since we have not only the seventeenth-century views by Speed and Hollar, but also three large contemporary plats of the fortifications east of the Hull River.[2] Thus, in spite of the fact that these works have long since disappeared, we have a fairly clear notion of their size and character. There were two large bulwarks located across the river from the northern and southern ends of the town, with a 'castle' lying approximately half-way between the two. Connecting the three fortresses was a crenellated wall almost 2,500 feet in length, which ran parallel to the river.[3] (See Pls. 7 and 8.) Reflection upon the character of these fortifications will suggest that they were designed to achieve the same goals that Henry's original plans had sought, but that they served other purposes as well. In the first place, by stretching along the eastern side of the Hull River, they provided solid defence against overland attack from the east, which neither the original scheme nor Henry's modifications had provided. On the other hand, the southern bulwark gave better coverage of the entrance to the haven than a bulwark in front of Watergate could have done, while the northern bulwark served the same purposes that had been intended for the 'platform' at North Gate. Thus both ends of the haven were strongly guarded against intrusion by enemy ships or infantry; yet if someone should break through, there was the heavily armed castle at the mid-point for them to contend with. This

[1] S.P. 1/169, f. 22.

[2] B.M., Cott. MSS., Aug. I. ii. 49; Aug. I, Supp. 3; Aug. I., Supp. 4.

[3] See Joseph H. Hirst, 'The Castle of Kingston-upon-Hull', *Transactions of the East Riding Antiquarian Society*, iii (1895), 24–39, for an historical, but undocumented, account of the fortifications.

latter point explains a curious fact about the two bulwarks and the castle, namely, that they had gunports on both the 'inner' and 'outer' sides of the ramparts. This was a clever way of using a single line of defence for protection against attackers who might come over land or water.

The best contemporary evidence for the Hull fortifications is supplied by the three large drawings mentioned above. One of these is partially reproduced in Plate 8, which shows only the left half of a plat that is nearly six feet and eight inches in width. In this reproduction the southern bulwark is seen on the left, the castle on the right, and the Hull River flows to the left and empties into the Humber. There is no scale provided on the drawing, although the length of the walls between the bulwarks and the castle, as well as the overall length of the fortification, is stated.[1] A second drawing (Aug. I. ii. 49) of the three fortresses is quite similar, with only minor variations in the placement of windows and doors. The one major difference is that it shows a moat some 50 feet wide which extends along the east side of the walls the full length of the fortifications. The drawing technique of the two plans is quite the same, except that the second is a little less precisely executed than the first. Oddly enough, the first drawing contains some dimensions but no scale, while the second displays a scale of 40 feet to the inch but no dimensions. A third drawing (Aug. I., Supp. 3) is even more roughly executed; it shows only the Hull River and the moat running parallel with it—neither the town nor the fortifications are indicated.

The drawing reproduced in Plate 8 gives the impression of being a 'show drawing'—a plat intended perhaps for the eyes of the king himself, to show the general character of the works. The fortresses and intervening walls are neatly drawn, the dimensions given in a formal handwriting, and the interesting but technically irrelevant row of merchants' houses shown along the west bank of the river. On the other hand, the second drawing may have been intended as a working drawing, since there is provided a scale which would be valuable to the workmen actually executing the design. Besides, there is the moat, which is a dominant feature on the plan and may very well have been the *raison d'être* for the drawing. This moat, by the way, was duly constructed, for it is clearly shown on Hollar's view of Hull. The third drawing—which shows only the river and the moat

[1] The overall length is given as 2,894 feet; from the castle to the bulwark next to the Humber is 1,066 feet; from the castle to the north bulwark is 1,400 feet.

and omits the fortifications as well as any dimensions or scale—may have been a preliminary sketch for the second drawing.

The three plans are closely related in drawing technique as well as subject-matter. All three are unusually large drawings and yet are executed on roughly the same scale. The type of paper used and the sepia-coloured ink are noticeably similar. Finally, there are similarities in drawing details, such as the circular scrawls to indicate water and the hatch-marks to show gunports. All of these similarities strongly suggest that the drawings were made by one and the same draughtsman. The documentary evidence connected with the Hull works is so meagre that there is little hope of determining on strictly documentary grounds who the draughtsman was. However, internal evidence from the drawings themselves suggests that they were drafted by Rogers. For instance, the similarity between the ground plans of the bulwarks at Guines and Hull would indicate that the man who designed the latter was thoroughly familiar with the former, and this description of course fits Rogers. Besides this similarity in design, the drawing technique of the Hull plats is reminiscent of those plats of Guines which we associated with Rogers while he was working there. Compare, for example, the gunports on the bulwarks of Guines Castle (Pls. 2 and 3) with those on the Hull fortifications (Pl. 8), where the shape of the ports and the use of hatch-marks show strong similarities.

It must be admitted, however, that even if we identify on the bases of design and of drawing technique the draughtsman of the Hull plats with the person who drew the plans and elevation of Guines, our argument for John Rogers is at best circumstantial, and at worst circular, for we have not categorically demonstrated in either case that he was that draughtsman. Fortunately, the inscriptions on the Hull drawing provide a more definite clue, for it can be shown that these inscriptions were written on the plan by Rogers's clerk who was working for him at Hull.[1] Probably it was the same clerk, mentioned earlier, who was assigned to help Rogers in his administrative position as comptroller and to whom we shall refer henceforth, for brevity's sake, as Clerk A. To be sure, this handwriting

[1] See Appendix A for an analysis of the handwriting of Rogers and his various clerks. Since this analysis constitutes a key factor in the arguments for Rogers's having drafted the various drawings under consideration in this study, the reader who is interested in the technical demonstrations of the arguments will be advised to read Appendix A at this point, before proceeding to the arguments in the text that are based on documentary, circumstantial, and stylistic evidence.

evidence still does not present categorical proof that Rogers executed the drawings, for while it is almost inconceivable that Rogers's clerk drafted the plans, still he could have penned the inscriptions on a plan drafted by someone other than Rogers. But as will be seen later, another set of drawings, executed after Rogers went to Boulogne, was inscribed by yet another of Rogers's clerks. While it is still possible that two of Rogers's clerks, at different times and different places, could have inscribed plans made by unknown draughtsmen, the likelihood seems rather remote, especially when it is remembered that Rogers himself was a well-known draughtsman, and that there is considerable documentary evidence that Rogers did draw plans at Boulogne. Circumstantial evidence thus points to Rogers as the draughtsman of the plans for the Hull fortifications, and the handwriting evidence points to Rogers's clerk as the penman who inscribed the explanatory words, phrases, and notes on the drawings. Furthermore, the handwriting evidence even offers an explanation for Rogers's use of his clerks in this manner, for his own writing was nondescript, and he apparently did not wish to spoil his formal drawings with his own informal script. Having reached these conclusions, we may also feel more assured that the Guines drawings—because of their stylistic similarity to the Hull drawings—were likewise by Rogers.

This conclusion regarding the draughtsman of the Hull plats suggests a hypothetical reconstruction of the circumstances in which the plats were made. When Rogers was called to Hull from Guines in 1541, he was no doubt informed of the resolutions for the Hull fortifications which Henry and his councillors had made on 4 October. Perhaps he recognized the weaknesses in the scheme proposed by the king, and so he formulated another plan for a much larger project across the river from Hull. The three plats of the fortifications would have been produced as Rogers worked out his ideas, while the more formal and finished of the three was probably intended for the king, in order to convince him of the merit of the new proposal.

Closely related to the drawings of the Hull fortifications is a large survey of the area around Hull (Pl. 9). Unfortunately, the drawing has been severely damaged, and a portion through the centre is missing.[1] This is

[1] The drawing has, I believe, been incorrectly pasted back together, for the portion missing is greater than that indicated by the blank space in the centre. This can best be detected by studying the size of the southern bulwark and the wall adjacent to it, both of which suggest that the blank space allotted for the length of the fortifications is inadequate.

doubly regrettable since the missing portion includes the larger part of Hull and the new fortifications. Yet enough remains to suggest the purpose of the drawing. The obvious area of interest is the upper right quadrant of the plan. Here the main concern seems to have been to show the drainage system of the fields. Although the vital centre portion of the drawing is missing, careful scrutiny of the remaining parts reveals that at least half of these drainage ditches led towards Hull and perhaps emptied into the moat outside the west wall of the town. The missing part of the drawing probably would have shown this clearly. If so, the survey fits in neatly with the royal resolution of October 1541, which called for 'the sluices to be viewed and new made that they may serve to drown about the town as the case shall require'.[1] Perhaps Henry VIII himself had ordered a graphic survey of the area around Hull to be executed, in order that the drainage system might be clearly understood.

It would have been quite natural for John Rogers, the king's surveyor at Hull, to have produced this survey of the area around the town. This inference is supported by the handwriting evidence, for the inscriptions on the survey were written by the same hand that inscribed the plan of the Hull fortifications. This of course means that the hand was not that of John Rogers, but rather of Clerk A. Thus we are again faced with the indirect argument that since it was Rogers's clerk who wrote the inscriptions, it was in all likelihood Rogers himself who drafted the plan.

If we are correct in assigning to Rogers the plans for the Hull fortifications and the survey of the area around Hull, then it becomes obvious that one of his more important duties as surveyor of the king's works was that of providing plats for the various projects. This point is further substantiated by another set of plans in the Cottonian collection which may be even more confidently attributed to Rogers, namely, four closely related drawings of Hull Manor. What we shall call Plat A sets forth the ground-plan of the building as well as the walls which enclose several gardens and the fields around the manor (Pl. 10).[2] Plat B (Pl. 11) appears to be the plan of the first floor of the house, while Plat C (Pl. 12) shows a similar layout for the first floor, although there are important differences in the two drawings. Plat D (Pl. 13) gives a bird's-eye view of the manor and of the walls and fields around the house.

[1] S.P. 1/167, f. 84.

[2] Only a portion of this large drawing is reproduced in the plate.

There are several reasons for supposing that these drawings were made by John Rogers. The endorsement on the verso of Plat C (Pl. 25) leaves little doubt regarding this particular drawing: 'A new plat made by the same Rogers, of the king his hyghnis manno^{r} of Hulle, the xxvth of June.' It seems quite safe to assume that this was the surveyor, John Rogers. The date would then have been 25 June 1542 or 1543.[1] Now the wording of this endorsement carries implications for the other drawings: 'A new plat' suggests that one or more plats of Hull Manor had been executed previously, while the phrase, 'made by the same Rogers', implies that Rogers had drafted the previous plat or plats. These inferences are supported by evidence from the drawings themselves. A detailed comparison of the drawings (to be made later) will show that Plats A, B, and D are closely related in content and that they precede Plat C in point of time of their execution. These conclusions harmonize with the implication in the endorsement that Plat C is 'a *new* plat'. Furthermore, all three floor-plans (A, B, and C) are produced on the same scale—roughly one inch for sixteen feet.[2] This suggests that all were produced by the same draughtsman, or at least that the draughtsman of Plat C had before him Plats A and B. Finally, although Plat C is a more highly finished drawing than either A or B, the basic similarities in the drawing technique of the three plans strongly suggest that they all came from the pen of the same draughtsman, namely, John Rogers. But apart from these arguments based on the close relationships between Plat C and Plats A and B, there is a point to note about the brief inscription in the centre of the floor-plan on Plat A, which reads 'voyd groūde'. This phrase was assuredly penned by Rogers himself (see Appendix A), and thus it constitutes independent evidence for associating him with the production of the drawing. Plat D, being a bird's-eye view, does not readily lend itself to stylistic comparisons with the floor-plans, but, as already noted, there is good reason for connecting it with Plats A and B, and thereby for assuming that it too was from Rogers's pen.

A more complicated question than that of the authorship of these

[1] Rogers remained at Hull from the autumn of 1541 until the autumn of 1543 before moving to a new assignment. There seems to be no sure way of deciding whether the plat was drawn in the summer of 1542 or 1543.

[2] No scale is stated on any of the plans; the above ratio was obtained simply by direct measurement, using the dimensions given on Plats A and B. Although the paper sizes of the three drawings vary considerably, the actual sizes of the floor-plans are the same. Unfortunately, this is not apparent in the plates, for the drawings have been photographed at different magnifications.

drawings is that of the purpose for which they were intended. A cursory survey might suggest that since Plats A and B were provided with dimensions they were building drawings to be used in actual construction work. Plats C and D—with their more careful draughtsmanship but without dimensions—would then be 'show drawings', intended only to inform the king of the work being done on his manor in Hull. But can it be safely asserted that *any* of these drawings were intended to be actual working plans? Henry VIII had a penchant for obtaining surveys of harbours, towns, castles, and houses, as the numerous drawings in the Cottonian collection demonstrate. Instead of being a set of building plans, the drawings could be merely graphic surveys of Hull Manor, similar to the survey which was made of Hull and the surrounding area.

A detailed analysis of the drawings might suggest a solution to the problem, but before this can be done adequately it will be necessary to gain some knowledge of the historical background of the building with which the drawings were concerned. Hull Manor—known at various times as Suffolk Palace, King's Manor, Hull Manor, Courthall—was built in the late fourteenth century by Michael de la Pole, at the same time that this energetic Duke of Suffolk was bringing to completion the city walls of Hull.[1] The manor remained in the hands of the de la Pole family until 1504, when it was confiscated by Henry VII at the fall of Edmund de la Pole, Duke of Suffolk.[2] The Crown retained possession of the manor until 1514, when Henry VIII granted the estate, along with other former possessions of the de la Poles, to Sir William Sydney, a young military figure who was rising in favour with the king.[3] Sydney kept the house until 1538, at which time he returned it to the Crown in exchange for lands in Sussex.[4] In that same year Henry ordered Leonard Beckwith, Receiver of Augmentations in Yorkshire, to make a survey of Hull Manor and the other properties that had been received back from Sydney.

When Henry visited Hull in September and October of 1541, he naturally stayed in his manor house and so had an opportunity to become personally acquainted with the building. Thus when he and his privy councillors formulated the plans for the fortification of Hull, Henry also included a resolution concerning the royal manor: 'Item, the king's majesty's

[1] Thomas Gent, *Annales Regioduni Hullini: Or, The History of the Royal and Beautiful Town of Kingston-upon-Hull* (London, 1735), 72.

[2] Ibid. 76.

[3] *L. & P.* i. 4909; cf. Gent, op. cit. 107.

[4] *L. & P.* xiv. i. 906 (7).

house to be made to serve as a citadel and a special keep of the whole town in such form as it shall please the king's majesty to devise and appoint the same.'[1] Largely on the basis of this resolution, historians of Hull have suggested that Henry 'beautified, repaired, and enlarged' Hull Manor, and it has been assumed that the drawings under consideration contain the changes which he made.[2] Our analysis of the purpose of the drawings will therefore require us to ascertain whether or not Henry actually did modify the building, and if so, what changes were made.

A crucial document for the resolution of these several problems is the Beckwith survey of 1538, which provides a fairly detailed description of Hull Manor before royal building activity began in 1541–2, and which therefore gives a point of reference for comparison with the drawings.[3] Leaving aside for the moment Beckwith's description of the gatehouse and entrance tower, we may turn directly to his account of the manor house: 'Also the court within the said tower contains half an acre of ground paved with stone and built round about with seventeen chambers of brick and stone and tiled. Whereof seven chambers is beneath and ten above with chimneys and jakes.' Comparing this description with the Hull Manor drawings, we note in connexion with Plat A (Pl. 10) that there are precisely seven rooms around the north court, if we begin with the first room to the right of the entrance way and move counter-clockwise. Counting in the same way, there are ten rooms shown on the first floor of Plat B (Pl. 11). On the other hand, the first floor as seen in Plat C (Pl. 12) shows some changes in room partitions on the east side of the court which make it difficult to count ten rooms in the same way that Beckwith seems to have done. Beckwith next described some of the rooms in the manor house: 'Also the hall contains in length sixty feet and in breadth forty feet built with brick and stone and covered with tile . . . also at the east of the hall is the buttery, pantry, and a chamber above them tiled.' Plat A shows this hall, with the buttery and pantry adjacent to it, exactly as Beckwith describes it. Likewise, Plat B has the chamber above the pantry, and a

[1] S.P. 1/167, f. 84.

[2] Gent, op. cit. 73 n.; followed by George Hadley, *A New and Complete History of the Town and County of the Town of Kingston-upon-Hull* (Kingston-upon-Hull, 1788), 700; John Tickell, *The History of the Town and County of Kingston upon Hull* (Hull, 1798), 186; and James J. Sheahan, *History of the Town and Port of Kingston-upon-Hull* (Beverley, 1866), 69.

[3] See Appendix B for a transcript of Beckwith's entire description of Hull Manor, from which the following extracts will be drawn for comparison with the plans. I am deeply grateful to Dr. D. F. Findlay of Hull for providing me with this transcript.

stairway from the hall to this chamber corresponds on the two plans. Careful scrutiny reveals that there is no entrance to the hall on the first floor of Plat B, which indicates that the hall was a two-storey room. Again, Plat C diverges considerably from the arrangement just described. A rather elaborate stairway has been added which provides an entrance to the first floor directly from the north court. A bay window and fireplace have been added to the chamber above the pantry, and doors have been opened into both the east and west ends of the hall, in addition to the one coming from the north court. It seems clear that another floor has been inserted, thus dividing the previous two-storey hall into upper and lower rooms.

We come next to Beckwith's description of the chapel. 'Also there is in the north side of the hall a chapel containing in length twenty-eight feet and in breadth fifteen feet, made of brick and stone and covered with lead, with a closet and sight into the chapel. Under the same is two chambers with a chimney.' This description corresponds closely with Plat B, where the chapel is clearly indicated by means of an altar table; furthermore, the 'closet' is seen to extend along the north wall of the chapel, just as Beckwith noted it. Plat C also shows the chapel, but the closet has been eliminated by removing the partitions in the adjacent room. When we move with Beckwith to the south court of the manor house, we find that his rather general description tallies with all three plans: 'Also there is on the south side of the hall another court containing a rod, built round about with stone and brick and covered with tile, with bakehouse, brewhouses, and other houses and lodging.'

From this comparison it has become clear that Plats A and B correspond rather well to those parts of the manor described by Beckwith. On the other hand, Plat C shows some marked differences from both Plat B and Beckwith's description. As for Plat D, the bird's-eye view of Hull Manor, it does not conform in all details to any one of the three floor-plans. There is one tell-tale clue in this drawing, however, that at least partially clarifies its relationship to the three floor-plans. Just above and behind the entrance tower can be seen the corner of a first-floor room which extends north of the hall, and, beyond this, the slight indication of a doorway which leads from the north court into the ground floor of the hall. These details conform precisely to Plats A and B, but they are completely out of correspondence with Plat C, which shows a large stairway at this place. Since Plat C diverges from the other three drawings on this important point, we may set it aside for the moment in an attempt to formulate more precisely the relationships between Plats A, B, and D.

A careful collation of Plats A and B reveals a fairly close correspondence between the architectural elements of the ground floor and the first floor as shown on these two plans. This can be seen most readily in the placement of internal stairs throughout the building and in the location of the privies on the first floor along the west wing of the house. But there are some disturbing discrepancies between the two drawings. For instance, there is an internal stairway in the east wing of the south court of Plat B which appears to be directly above the fireplace of the room below as shown on Plat A. And of the four external stairs on Plat B—two in the south court, one in the north court, and one on the northernmost wall of the house—only one is shown on Plat A. Finally, the numerous bay windows on Plat B are nowhere to be seen on Plat A. These important differences make it difficult to suppose that both plats could be surveys of the building as it stood at the time they were drawn, for in that case they would have corresponded more precisely. Comparison of the two plans with Plat D suggests a solution to the problem. The bird's-eye view is obviously more closely related to Plat B than to Plat A, for it shows both the bay windows and the external stairs that are prominent features of Plat B. But further observation reveals that the bay windows do not correspond even on these two plats, nor do the locations of the chimneys and fireplaces. Thus both B and D cannot be exact surveys of an existent building.

From these considerations a hypothesis begins to emerge, namely, that Plat A is a survey of the ground floor and garden walls as they existed at the time the drawing was made, whereas Plats B and D show modifications in the building through the addition of bay windows, fireplaces, and external stairs. The addition of these three elements to the first floor of the building was clearly intended to make the rooms brighter, warmer, and more accessible from the courtyards of the house. In short, Plats B and D represent proposals for remodelling the first floor of the manor house in order to make it more comfortable for the royal presence. Just why they do not correspond more closely it is difficult to say, particularly if it is assumed that they were drawn in conjunction with one another for the purpose of presenting to the king the proposed remodelling. But since they do disagree in several instances on the location of precisely those elements involved in the remodelling—namely, bay windows and fireplaces—one must suppose either that Rogers was careless in correlating these elements on the two drawings, or else that he changed his mind regarding their placement. The latter alternative seems the more likely; if it is true, then one of the drawings represents a later stage in his thinking than the other, although there seems to be no way of deciding which came first.

One other point about Plat B deserves attention. The close correlation between this drawing and the description of the rooms in Beckwith's survey of 1538 has already been noted, so that Plat B probably shows few, if any, proposals to change partitions or arrangements of the rooms. However, if our hypothesis is correct that several of the stairs represent merely *proposed* additions rather than existing elements, then it may be surmised that some of the interior doors of the first floor of the actual building are not shown on Plat B, for without the external stairs there would have been no means of access to these apartments. This combination of new stairs with the walling up of old doors suggests that Rogers was trying to redistribute the flow of traffic on this first floor to accommodate the structure to the habits, customs, and needs of its royal occupants. To this end he devised five apartments around the two courts, each apartment being sealed off from its neighbours and provided with an internal or external stairway or both. The largest block of rooms surrounded three sides of the north court and was clearly intended for the king. But this arrangement hardly left a satisfactory apartment for the queen, nor indeed was it completely suitable for the king.

Whether it was Rogers or Henry VIII or both who felt dissatisfied with the scheme for remodelling shown on Plats B and D, we do not know. At any rate, 'the same Rogers' returned to his drawing board and produced 'a new plat'. This time there emerged a striking renovation of the first floor-plan (Plat C) which could fully accommodate the courtly, ceremonial, and physical needs of both the king and the queen. Unfortunately, Plat C has no inscriptions indicating the purpose intended for the various rooms, nor is there any direct documentary evidence relating to the drawing. Any attempt to designate the functions of the many rooms on this plat must perforce be highly conjectural, but there seems to be no harm in trying, and indeed a better grasp of the character of the drawing might result from the effort.

The general layout of the floor-plan suggests that Rogers was striving for an internal arrangement similar to the royal apartments which Henry had recently caused to be renovated or built afresh at Windsor Castle, Hampton Court, and Nonsuch Palace.[1] None of these has survived intact today, but their general scheme can be reconstructed from documentary, pictorial, and archaeological data.[2] The apartments consisted of a series of adjoining chambers through which persons seeking audience with the king had to pass. Audience might be granted in any one of several chambers, but the further the chamber from the entrance to the apartment, the more private the audience. Thus the 'state rooms' led by degrees into the more private quarters of the apartment, in the deepest recesses of which was located the king's bedroom. These apartments were placed on the first floor of the building, with the ground floor providing storage space, kitchens, workshops, and quarters for the numerous servants and officials in attendance upon the king and queen. It was a Tudor practice to arrange these elements around two courts, as Wolsey had earlier done at Hampton Court and Henry was currently doing at Nonsuch Palace. With its two courts Hull Manor thus lent itself readily to royal accommodation. By keeping in mind contemporary trends in English royal palace-building, we can make an educated guess at Rogers's intentions for Hull Manor as shown on Plat C.

One of the most significant changes involves the great hall which separated the two courts. Two doorways leading from the hall into rooms on the first floor plainly indicate that another floor has been inserted into the two-storey hall, dividing it into upper and lower rooms. This was not an uncommon Tudor practice in remodelling older houses.[3] Although a splendid two-storey hall was built for Henry at Hampton Court, at Nonsuch this feature was entirely suppressed in favour of a series of state rooms. The latter alternative appears to have been the intention for Hull Manor as shown on Rogers's drawing. To give access to this upper chamber,

[1] This point was first suggested to me by Mr. Martin Biddle, who recognized the character of Plat C as a result of his own excavations of Nonsuch Palace. I am happy to give him credit for this insight, though of course he should not be held responsible for any errors which I may make in the following analysis.

[2] See W. H. St. John Hope, *Windsor Castle* (London, 1913), i. 252–6 and ii. 569, fig. 86; Ernest Law, *The History of Hampton Court Palace* (London, 1885), i. 176, 202, and double-page plan prefixed to p. 1; and for Nonsuch, Martin Biddle, 'Nonsuch 1959–60: An Interim Report', *Surrey Archaeological Collections*, lviii (1961), 1–20; idem, 'Nonsuch Palace', *History Today*, xi (1961), 206–14; and John Dent, *The Quest for Nonsuch* (London, 1962), 102–5.

[3] T. Hudson Turner and J. H. Parker, *Some Account of Domestic Architecture in England, from the Conquest to Henry VIII* (Oxford, 1859), iii. 55; cf. Margaret Wood, *The English Mediaeval House* (London, 1965), 49 and 61.

Rogers designed a grand staircase leading from the north court. (See Fig. 2, *A* and *B*.)[1] In the room to the east of the hall he inserted a fireplace and bay window; one might suppose these conveniences were intended for the council chamber (*C*).[2]

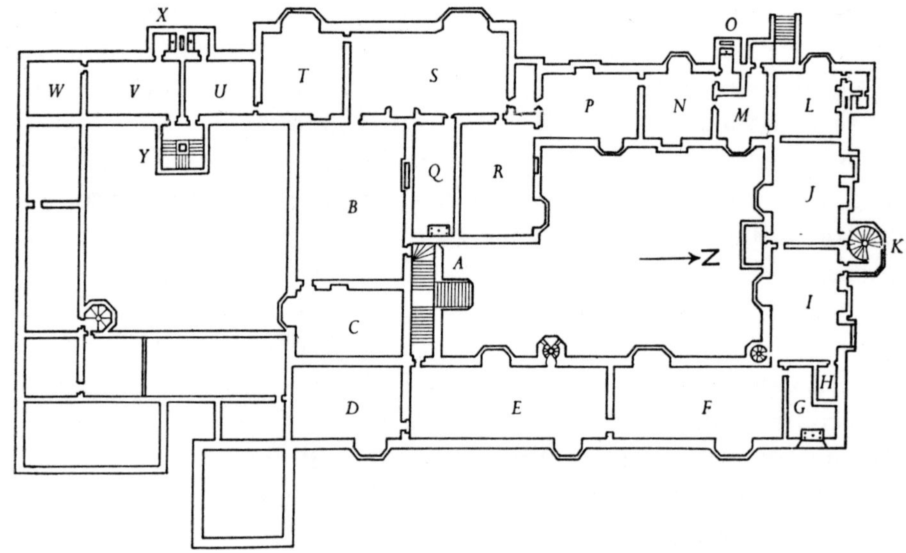

FIG. 2. *Conjectural Location of Chambers on Plat C of Hull Manor*

A. Grand Staircase
B. Hall
C. Council Chamber
D. King's Watching Chamber
E. King's Chamber of Estate
F. King's Presence Chamber
G. King's Chapel
H. King's Chapel Chamber
I. King's Privy Chamber
J. King's Dining Chamber
K. King's Privy Stair
L. King's Pallet Chamber
M. King's Closet
N. King's Bedchamber
O. King's Stool Chamber
P. King's Great Chamber
Q. Queen's Chapel
R. Queen's Watching Chamber
S. Queen's Presence Chamber
T. Queen's Privy Chamber
U. Queen's Pallet Chamber
V. Queen's Bedchamber
W. Queen's Closet
X. Queen's Stool Chamber
Y. Queen's Privy Stair

The grand staircase, besides giving access to the hall, also led to the east range of the north court. The three large rooms of this range, each with one or two bay windows, could have served as the king's watching chamber or guard room (*D*), the chamber of estate or first waiting room (*E*),

[1] For the next couple of paragraphs the letters in parentheses will refer to this figure.

[2] At least, there was a similar arrangement of hall and council chamber at Windsor Castle.

and the chamber of presence where the king gave formal audience (*F*). This procession of state rooms was interrupted at the north-east corner by the location of the king's chapel (*G*), which contained a chapel chamber where the king might observe the service in private (*H*).[1] Around the corner from the chapel lay a room with a fireplace and two bay windows. This might have been intended for the king's privy chamber (*I*), and beyond this, perhaps, the king's dining chamber (*J*). The large well-stair leading into both of these rooms could have served as the king's privy stair (*K*). At the north-west corner of the house was a room suitable for the king's pallet chamber where the officers of the bedchamber would sleep (*L*). Adjacent to this was a small room, perhaps the king's closet (*M*), and beyond this the king's bedchamber (*N*) with a bay window, fireplace, and adjacent stool chamber or garderobe (*O*). The next room, likewise provided with light and heat, might have been intended as the king's great chamber (*P*).

If our conjectures are anywhere near to what Rogers had in mind while redesigning the first-floor plan of Hull Manor, it will be seen that the king's apartments absorbed most of the area around the north court. For the queen to have received adequate accommodation, her apartment would have had to extend into the south court. In fact, the proper rooms could be rather nicely fitted into the space along the east range of the house. There was a chapel for the use of the queen (*Q*), and three 'state rooms' would include the queen's watching chamber (*R*), presence chamber (*S*), and privy chamber (*T*), each provided with a bay window and fireplace to improve the comfort of the rooms. It should be noted that in order to obtain a satisfactory privy chamber for the queen, Rogers had to eliminate an internal stairway and a garderobe that are shown on Plat B. Next to the privy chamber was a room with garderobe that could serve nicely as the queen's pallet chamber (*U*). Beyond this lay the queen's bedchamber (*V*), with her own closet (*W*) and stool chamber (*X*). Access to the apartment from the south court was provided by what might have been the queen's privy stair (*Y*).

Finally, an apartment with five chambers was located on the south and east ranges of the south court. Access to the apartment was by means of

[1] This too was a common practice in regal and lordly homes in that day. Plat B clearly shows a window in the chapel chamber which overlooks the altar of the chapel, but for some reason Rogers has not shown this window on Plat C. He may have forgotten to draw it, or he may have intentionally eliminated it.

a well-stair leading up from the court. That this apartment was intended for lesser dignitaries is indicated by the fact that none of the rooms was provided with a bay window, fireplace, or garderobe.

Rogers's 'new plat' presents a rationalized plan for modernizing Hull Manor along the lines of the latest royal apartments at Windsor, Hampton, and Nonsuch. No doubt he had had opportunity to acquaint himself with these arrangements while working at Hampton Court in 1533–7.[1] Since Nonsuch Palace was begun in 1538, he could possibly have worked there also before being called to Guines in 1541.[2] At any rate, this interesting proposal for renovation suggests that Rogers might in time have acquired a reputation as an architect, had his career in the Office of Works not led him to specialize in military and civil engineering.

Perhaps it will be well to summarize the results achieved thus far in our analysis of the drawings of Hull Manor. It is fairly certain that all four were drafted by John Rogers. The first to be drawn was Plat A, which appears to be a graphic survey of the ground floor and garden walls as they stood at the time the drawing was made. Plat B, a plan of the first floor, and Plat D, a bird's-eye view of the house and grounds, were perhaps made about the same time. Conforming fairly well to Plat A, they are more than surveys, for they contain proposals for remodelling the first-floor apartments of the house, although the proposals on the two drawings do not exactly correspond. Plat C is a later plan for a more drastic remodelling of the first-floor apartments.

Authorship, purpose, and proper chronological sequence of the drawings having now been established, the question remains as to whether or not any of the changes proposed in the drawings were actually executed. If there were any changes, the building accounts for the work unfortunately have not survived. So the answer to the question will have to be based on fragmentary evidence drawn from incidental references in contemporary documents, from the later views of Hull by Speed and Hollar, and from the character of the drawings themselves. On one point, however, the Beckwith survey serves us again very nicely. In 1538 Beckwith described the entrance to the manor house as follows:

[1] It will be recalled that while there Rogers was paid for cutting beckets for the door of the king's great watching chamber. See above, p. 6, n. 2.

[2] Unfortunately, the one surviving particular account for Nonsuch during its construction (P.R.O., E. 101/477/12) covers only the first work season, 22 April to 14 September 1538. The lists of masons in the account do not include Rogers's name.

That is to say, in the first entry into the same mansion place out of the said street called Market Gate, there is first a gatehouse built with timber and covered with tile, wherein is two chambers. And the entry is built on both sides with warehouses of timber work, covered with tile, containing in length an hundred feet and in breadth twenty feet.

Plat A fails to show either the gatehouse or the warehouses, but instead indicates merely a straight wall with a simple entrance leading into an empty forecourt east of the house. It appears that between the time of the survey of 1538 and the drafting of Plat A the wooden gatehouse and warehouses were taken down, leaving the interior of the forecourt unobstructed. This was very likely the result of Henry's desire to make Hull Manor into a 'citadel', as stated in the resolutions of October 1541. With this point in mind we may note in particular the stirrup-shaped bastion inserted into the wall of the forecourt as shown on Plat D. Placed low in the bastion are three gunports, two facing in opposite directions along Market Gate, a third looking directly down Chapel Lane. It seems possible that some such defensive work was constructed, for Speed's map of 1610 clearly shows a semicircular protuberance on the forecourt wall precisely opposite the juncture of these two streets.[1] The other defensive measures which Rogers has shown on Plat D are almost equally interesting, for he has placed low gunports in the angles of the exterior garden walls in such a way that gunfire can be directed down all the streets parallel to or approaching the grounds of the manor house. Whether or not these gunports were actually inserted into the garden walls remains an open question, for the views by Speed and Hollar are too small in scale to show such details.

Regarding the proposals for remodelling the house itself as shown on Plats B, C, and D, there is scant evidence that any such proposals were actually carried through. Certainly one important modification shown on Plat C was not executed. In the room which we have conjecturally designated as the chamber of estate (Fig. 2, *E*), Rogers has removed a partition which was a load-bearing wall supporting the south face of the tower above the room. To have removed this partition either the tower would have had to go or else some other means of support would have had

[1] Speed, loc. cit. Admittedly, this element is lacking on Hollar's drawing (Pl. 7), but it might be suggested that from the point of view which Hollar has taken, the bastion is blocked from sight by the tower of the house.

to be inserted into the room. On the one hand Plat C shows no such support, and on the other hand Speed's and Hollar's seventeenth-century views show the gate-tower as still a prominent feature of the house. Hollar's view, though it lacks sufficient detail, further suggests that the other main elements of the proposals for remodelling—namely, bay windows, well-stairs, and fireplaces—were not added to the house at this time. Certainly the large well-stair on the north face of the house, which constitutes a prominent feature of Plats B, C, and D, is not to be seen on Hollar's drawing, although it would have been easy enough to show it.

The absence of building accounts, the lack of contemporary references to building works at Hull Manor, and the later views by Speed and Hollar all point to the conclusion that little was done to remodel the house according to the drawings made by Rogers. Furthermore, the rather casual attitudes of Henry VIII and Edward VI towards Hull Manor indicate that no great amount of royal treasure or energy had been devoted to remodelling. In February 1544 Henry allowed Laurence Wharton 'to have the keeping of the King's manor . . . with 6*d.* a day out of the revenues of Beverley Manor'.[1] Edward VI granted Hull Manor to Walter Jobson in March 1550 for a payment of £277. 2*s.* 8*d.*[2] Finally, in March 1552 Edward granted the manor to the mayor and burgesses of Hull for an annual payment of £13. 0*s.* 4*d.*[3] It is thus quite likely that Hull Manor passed out of the royal hands in 1550 in much the same condition that it had been received in 1538. For the next century the building underwent a chequered fate; utilized for a while as a market, it was rented in 1658 at £60 per annum for use as a magazine. Shortly thereafter the property was sold in separate lots and the house itself was pulled down.[4]

Although the progress of the fortification projects in Hull cannot be traced in detail, nor can Rogers's specific activities as surveyor be described, there is sufficient reason to believe that the king was satisfied with the works, and that his esteem for Rogers was continually rising. Regarding the latter point, Henry was beginning to use Rogers as a technical consultant for royal projects that were not directly under Rogers's supervision. In April of 1542 the king instructed his surveyor to leave the works

[1] *L. & P.* xix. i. 141 (66).

[2] *C.P.R., 1550–3*, 5.

[3] Ibid. 334. John Travis-Cook, *Notes Relative to the Manor of Myton* (Hull, 1890), 133 ff., tries to explain the startling drop in value of the manor as reflected in this payment.

[4] L. M. Stanewell, ed., *Calendar of the Ancient Deeds, Letters, Miscellaneous Old Documents, etc., in the Archives of the Corporation*, (Hull, 1951), 151 (D. 943, d–k); cf. Gent, op. cit. 73 n.

at Hull in charge of overseers and go to Berwick-on-Tweed to view the fortifications. Since, as Henry noted, Rogers knew the plats of these fortifications, he could judge whether they corresponded to the actual works. In his review of the works Rogers was to note any faults which needed to be corrected and to give any other advice that seemed necessary.[1] In June of the same year Rogers again found himself in the role of technical consultant, when he was sent to Wark Castle, Northumberland, to view the works and report on how the castle might be strengthened. Furthermore, he was to make plats of the castle 'according to his instructions'; presumably the instructions were that the plats should contain his suggestions for improvements. If these plats were ever made by Rogers, they seem not to have survived; nor is Rogers's written report on the condition of the castle extant.[2]

Another indication of Rogers's increasing favour with Henry was the annuity of £36. 10*s.* 0*d.* which was granted on 1 January 1543 to 'John Rogers, the King's servant . . . in recompense of the fee which he formerly had of the king'.[3] The significance of this sum becomes apparent when it is compared with the annual fees of the king's master masons, who often received less than half the amount of Rogers's annuity.[4] Equally striking is the fact that Richard Lee, Rogers's immediate superior at Guines in 1541, had been appointed in 1536 as surveyor of Calais at a fee of £30 a year, plus a house to live in.[5] On the other hand, though Rogers had achieved a high economic status relative to the master masons and other surveyors in the king's service, his pay was still far below that of the special consultant engineer from Germany, Stefan von Haschenperg, who in 1540 received an annuity of £75.[6]

When Rogers finished his work as a technical consultant at Wark Castle, he returned to Hull to continue his supervision of the fortification of the

[1] *L. & P.* xvii. 277.

[2] The report which Rogers made is mentioned in a letter of September 1542: ibid. 726. Incidentally, the works accounts for Wark Castle have survived for the period 12 February 1542 to 10 November 1543 (B.M., Harl. MS. 1724), and Rogers is nowhere mentioned in them. His expenses in going to Wark must have been handled in a different account.

[3] Ibid. xviii. i. 100 (3). Although the identification is not clearly stated in the grant, this John Rogers is most likely the one with whom we are concerned. In a letter of 9 December 1543 the phrase 'the king's servant' is applied to Rogers the surveyor (ibid. xviii. ii. 473). This was not an uncommon appellation even for technical officials at that time; in 1546 the master mason at Calais, John Wright, was so called (ibid. xxi. i. 302 (3)).

[4] Cf. the various remuneration tables in John Harvey, *English Mediaeval Architects* (London, 1954), 328–32.

[5] *L. & P.* xi. 385 (13).

[6] Ibid. xv. 942 (108).

town. There is little that can be said about his precise activities or the progress of the works. In April 1543 Michael Stanhope, lieutenant of the garrison, declared the amounts of money that had been received and spent on both the garrison and fortification works. The large amount of the latter suggests that a considerable labour force was being employed.[1] An allusion to the 'perfection of the works at Hull' which is found in a letter dating from November 1543 suggests that the fortifications were nearing completion by that date.[2] Since these works were begun in the autumn of 1541 there would have been sufficient time by the autumn of 1543 to complete even the large project east of the Hull River.

Realizing that the work at Hull would soon be completed, Henry was already making plans for his surveyor. In the letter just mentioned he informed administrative officials at Wark Castle and in the city of Carlisle that Rogers would soon be moving to these works. The letter itself does not specify what Rogers's official position would be, and indeed it implies that the point had not yet been decided, but would be in the near future. It seems that there had been some difficulties in the progress of the works at Carlisle. Stefan von Haschenperg, the German engineer who was the technical supervisor of the works, had first been called to Court to account for his work and had then been relieved of his duties.[3] Perhaps Henry's intention was to send Rogers to Carlisle in order to clear up the difficulties in the royal works in that town, although it is difficult to say in what capacity he would have gone. There are two contemporary drawings in the British Museum, one showing the entire city of Carlisle, the other being a plan of Carlisle Castle, which reveal some similarity to the drafting style of the other drawings attributed to Rogers.[4] But the indecisiveness of these features, when combined with the lack of documentary evidence that Rogers drew plats of Carlisle or even that he ever worked in the town, makes the attribution of these drawings to Rogers a very dubious one.

Even if Rogers was working at Carlisle and making plats of the town and castle, this was certainly not his only occupation, for Henry was using him at the same time as a consultant military engineer in the war against Scotland. Rogers's technical knowledge of fortifications and his drafting

[1] £14,851 for the period 15 February 1542 to 21 April 1543: *L. & P.* xviii. i. 433.

[2] Ibid. xviii. ii. 395.

[3] Ibid. xviii. i. 901; cf. B. H. St. J. O'Neil, 'Stefan von Haschenperg, an Engineer to King Henry VIII, and His Work', *Archaeologia*, xci (1945), 147.

[4] Cott. MSS., Aug. I. i. 11 and 13.

skill made him particularly valuable in spying out enemy positions and in drawing plats of the enemy's defences. We have seen him performing this work while he was master mason at Guines, and we shall see him at it again when he becomes surveyor at Boulogne. Meanwhile, in December of 1543 Henry informed the Duke of Suffolk, who was commanding the English forces against the Scots, that he was sending Rogers to Tantallon Castle, East Lothian, 'to viewe the castle secretely and to bring his majeste a true plat and perfaict description of every parte of the same'.[1] Tantallon was the ancestral home of Archibald Douglas, Earl of Angus, the powerful Scottish noble who had been mediating the difficulties between the Scots and the English, but who, Henry feared, was now scheming to side with his fellow countrymen. Rogers's task was to gather technical information on the defences of Tantallon which could be used in a siege of the castle if Angus did revolt. This information would be presented in the plat which Henry requested. If that plat was ever drafted, I have found no indication that it is still extant.

Henry must have been pleased with the espionage work of his surveyor, for by the end of May 1544 plans were being laid for Rogers to make another trip into Scotland, this time in the company of three other men. Orders were sent to mayors and sheriffs along the route to provide horses for the men, and to the Earl of Hertford, lieutenant-general in the north, to furnish a ship which could secretly transport the men to the mouth of the River Tay.[2] On 1 June Hertford reported from Darlington, Co. Durham, that the ship was ready to depart as soon as John Rogers and the others should arrive there; by 17 June the journey into Scotland had been completed.[3] The purpose of this sojourn remains obscure, for the whole affair was executed under considerable secrecy. At one point Hertford was informed that the men were going into Scotland 'for suche purposes and to suche ende as by the instructions which you shall receyve herwith may more att large appere unto yow'.[4] But these instructions, if they were in written form, seem not to have survived. Whatever the mission was, Rogers must have played a major role in it, for it was he who reported in person to the king on the results of the journey.[5]

Meanwhile, Rogers had been appointed to a post which takes us

[1] Joseph Bain, ed., *The Hamilton Papers* (Edinburgh, 1892), ii. 219.

[2] *L. & P.* xix. i. 589, 594.

[3] Bain, op. cit. ii. 397, 413.

[4] Ibid. 394.

[5] Ibid. 413.

somewhat by surprise. On 28 February 1544 John Rogers, 'the King's servant', was granted the office of Clerk of the Ordnance in England to replace William Huxley, who had recently died.[1] There is no doubt that this was the same Rogers as the surveyor of works at Hull, for he rendered account as Clerk in the following July. This account has survived, and the signature which is affixed to it corresponds precisely with the signatures on letters written by Rogers the surveyor.[2] Nevertheless, there are still some puzzling features about this appointment. For instance, it is not clear why a building technician like Rogers was assigned to the position of Clerk of the Ordnance. To be sure, Rogers's activities since 1541 had been devoted to military matters in one way or another, so that his appointment to the Office of Ordnance in 1544 might not appear entirely unrelated to his recent activities. But why the *Clerk* of the Ordnance? Henry VIII had just reorganized the Ordnance Office in 1543 and had created several new posts, including Lieutenant of the Ordnance and Surveyor of the Ordnance.[3] One of these offices would seem to have been more suitable for Rogers. A possible explanation for the appointment is that Rogers already held a position as a lesser clerk in the Ordnance Office, and that the grant in 1544 represents a promotion to the higher office of *the* Clerk of the Ordnance. This hypothesis is based upon a seventeenth-century document which lists Rogers as one of several 'Clarkes in tempore King Henry ye Eight within Thoffice of his Majesties Ordynance'.[4] Here Rogers is credited with receiving a salary of 6*d. per diem*. Later in this same document (fol. 32ᵛ) Rogers is listed as 'Clarke of Thordnance' at 8*d. per diem*, which was the salary stipulated in his official appointment to the office on 28 February 1544. The implication that this salary increase represents a promotion within the Ordnance Office should not, however, be overemphasized. After all, this document was composed a century after Rogers's appointment and should therefore be used with caution.[5] Further-

[1] *L. & P.* xix. i. 141 (72).

[2] B.M., Add. MS. 5753, f. 121ᵛ. See Appendix A and Pl. 26.

[3] O. F. G. Hogg, *The Royal Arsenal: Its Background, Origin, and Subsequent History* (London, 1963), i. 36.

[4] B.M., Add. MS. 6176, f. 29.

[5] Fols. 29–39ᵛ list the various officers of the Ordnance from *c.* 1525 to *c.* 1640, with the years when particular incumbents were granted the offices, in addition to examples of expenses entailed in each office and the duties and responsibilities of the various officers. On fol. 35 there is a detailed analysis of the duties and privileges of the Clerk of the Ordnance and the Clerk of Deliveries, which suggests that this whole compilation may have derived out of those interminable debates between the officers of the Ordnance Office which were characteristic of the late Tudor period (cf. Hogg, op. cit. i. 45–53). If this is the case, then confidence in the document is even further weakened.

more, it is hard to visualize Rogers as a scribal clerk within the Ordnance Office, when we recall his background as a working mason, and especially when we consider the character of his handwriting, which was not that of a trained scribe.[1] Certainly we may reject the assertion that Rogers had been a clerk under the Master of Ordnance as early as 1523.[2]

If the questions as to how and why Rogers became Clerk of the Ordnance remain open, there are likewise some puzzling features about his tenure of that office. Except for the few months immediately following his appointment, Rogers was quite inactive as Clerk of the Ordnance for the thirteen years that he held the office. Yet it does not seem likely that this office was intended to be a sinecure: Henry VIII was greatly interested in developing the Ordnance Office, so that he would hardly have expected the important post of Clerk of the Ordnance to become dormant. Besides, the account which Rogers rendered as Clerk in July 1544 indicates that in the first months of his tenure he was busily performing the duties implied in the office: we see him supervising the shipment of ordnance between various points in England and at Calais, travelling back and forth between Westminster Palace and the Tower of London to check on the ordnance stores, paying wages to wagoners for transportation of ordnance supplies, and of course rendering account for the expenses involved in his office.[3] But that account is the last we hear of his activity as Clerk of the Ordnance. He held the position for the rest of his life, but I have found no evidence in the state papers that he actively functioned as the Clerk. Instead, the duties which he might have been expected to perform were fulfilled by Sir Francis Fleming, Lieutenant of the Ordnance, or Anthony Anthony, Surveyor of the Ordnance.

The explanation for Rogers's inactivity as Clerk during Henry's reign appears to be that he was so involved in other services to the Crown that he was allowed to let fellow officers of the Ordnance perform his duties. But even in the later years of his life, when he was no longer deeply committed to major royal projects, Rogers did not resume the duties of the Clerkship.[4]

[1] See Appendix A.

[2] A. H. W. Robinson, *Marine Cartography in Britain* (Leicester, 1962), 18. This assertion is based on a misreading of fol. 29, where the date written at the top of the page ('anno 14 Henry VIII', i.e. 1523) is mistakenly applied to the indefinite phrase 'Clarkes in tempore King Henry ye Eight', under which rubric Rogers is listed.

[3] Add. MS. 5753, ff. 161^{v}–21.

[4] Although the office did remain dormant as long as Rogers lived, it was revived upon his death with the appointment of William Painter, who functioned as an active Clerk of the Ordnance for more than thirty years. The difference in the activities of these two Clerks was noted by the compiler of Add. MS. 6176, who remarked

It had been in the autumn of 1544, however, that his activity as Clerk of the Ordnance had waned, when he was appointed surveyor of the works at Boulogne. This latter position kept him busily engaged for almost six years; during this period he reached the high point of his career as a military engineer.

that 'apon the death of Huxley one Rogers procured a pattent thereof to himselfe pro termino vitae with the ffee of viii[d] per diem but dealt nothing at all with thoffice of store' (fol. 35). On the other hand, he lists on fol. 32[v] some of the many payments to Painter for expenses connected with the office of Clerk of Ordnance—travelling, paying bills, and keeping account books.

IV

THE ENTERPRISE OF BOULOGNE

THE war between France and England which had been threatening for several years finally emerged in the summer of 1544. Henry capitalized on the current difficulties between Francis I and Charles V by forming a war alliance with the emperor against the French king. Henry and Charles planned a simultaneous attack on France, the imperial forces to drive in from the east, the English forces from the north. According to the plan of this 'Enterprise of Paris', the two armies would meet on the Marne and make a combined assault on the French capital.[1] Henry's initial task in his part of the project was to secure Picardy by capturing the two fortress towns of Boulogne and Montreuil, before pushing south towards Paris. Charles had a similar task in taking Luxemburg and St. Dizier. Though Luxemburg succumbed quickly enough, the defenders of St. Dizier proved obstinate and held up the imperial forces two months past schedule. Henry encountered difficulties at both Boulogne and Montreuil. Using Calais as a springboard for operations, he personally directed the siege of Boulogne, while the Duke of Norfolk handled that of Montreuil several miles to the south. Both towns were well fortified and Montreuil at least was strongly garrisoned. It took Henry six weeks to bring the relatively small garrison at Boulogne to its knees on 14 September; Montreuil never fell to the English.

Meanwhile both Charles and Henry were losing interest in pushing the 'Enterprise of Paris'. On 18 September at Crepy Charles even proceeded to make peace with the French without consulting his English ally. Left in the lurch, Henry, who had captured Boulogne only four days previously, was determined to keep the town as the sole reward of his expensive war. But he soon found that keeping his trophy might prove even more expensive than getting it, for with Charles V out of the war, the entire French army could be directed to the task of recapturing Boulogne. Having been the besieger of the town, Henry now became its defender. Mere force of

[1] Charles Oman, *A History of the Art of War in the Sixteenth Century* (London, 1937), 337–8.

English arms was, however, insufficient; it seemed desirable to improve the older French defences in order to guarantee the security of the town against the inevitable attempt of the French to regain it. But when Henry embarked upon a programme of refortification at Boulogne, he got himself involved in a vicious circle of expenditure: the more he spent on fortifying and garrisoning the town, the more imperative it became to hold on to it as a justification for the money already spent, and therefore the more necessary it was to keep spending money so that he would not lose his prize back to the French, who were very anxious to recover it.

John Rogers was to play a major role in the work at Boulogne, for as the surveyor of works he was not only the technical supervisor of construction, but also an administrator directly involved in spending and accounting for money spent on the works. However, he did not participate in this 'Enterprise of Boulogne' from its very beginning. The technical aspects of the siege operations at both Boulogne and Montreuil were under the supervision of an Italian artist and engineer by the name of Girolamo da Treviso, or, as the English called him, Jeronimo.[1] Henry seems to have had some doubts about this man. Perhaps he was becoming dubious of foreign engineers in general—we have noted his unhappy experiences with the Portuguese engineer at Guines and with Stefan von Haschenperg at Carlisle. At any rate, Henry asked for an opinion on Girolamo from Lord Russell, who was helping to command the siege at Montreuil. Russell replied that he found the man willing enough, but rather inexperienced in sieges.[2] This judgement was perhaps ironically borne out when Girolamo was killed on 11 September in an assault on the castle of Boulogne, after he had directed the mining operation which exploded part of the castle wall.[3] Three days later the town surrendered, and the English immediately faced the necessity of reconstructing the battered walls and of building additional fortifications in preparation for the *revanche* of the French. For this task Henry turned to his English engineers.

At first, however, Henry himself took charge of planning the work. On 18 September he entered the town and for the next several days occupied

[1] On the English phase of Girolamo's career, see R. W. Carden, 'The Italian Artists in England during the Sixteenth Century', *Proceedings of the Society of Antiquaries*, 2nd ser., xxiv (1911–12), 199–202; A. E. Popham, 'Hans Holbein's Italian Contemporaries', *Burlington Magazine*, lxxxiv (1944), 13–14; and Philip Pouncey, 'Girolamo da Treviso in the Service of Henry VIII', ibid. xcv (1953), 208–11.

[2] *L. & P.* xix. i. 1005.

[3] Ibid. xix. ii. 216.

himself with viewing the state of the fortifications, planning the works, and organizing both the military and civilian personnel who would carry through his plans.[1] It was this period of personal observation, by the way, that gave Henry the intimate knowledge of the requirements for the town's proper defence which he was later to display in his correspondence and discussions with the officers in charge at Boulogne. By the time Henry was ready to depart for England on 30 September, he had filled a whole roster of military and civilian posts.[2] John Dudley Lord Lisle was appointed the military commander of the town. This important post was to be filled with a series of officers during the remainder of the English occupation. In January 1545 Lisle was replaced by Sir Thomas Poynings, who was himself replaced by William Lord Grey in April 1546. Grey lasted somewhat more than a year until the position was taken over by Sir John Brydges in August of 1547. By May 1549 Edward Lord Clinton had replaced Brydges. This parade of officers in the top military post contrasts with the much greater stability in the top civilian post—that of surveyor of the works. In Henry's initial organization in September John Rogers was named 'surveyor of all the fortifications and works', while a certain Reynolds was appointed as master mason, Batte as master carpenter, Philips as chief smith, and Barwel as clerk of the works. A month later part of Rogers's duties were removed from him when the works at the entrance to Boulogne haven were placed under the charge of Thomas Palmer, who had been the treasurer at Guines Castle, and who apparently was well-versed in military architecture and building. Indeed, Palmer was merely one of several technical advisers who were called in at this time to assist Rogers in planning the work. Richard Cavendish, who possessed technical knowledge in building, was made the comptroller of the works, while Richard Lee (now Sir Richard), formerly surveyor of works at Calais, and John Burgate, at present the surveyor there, were brought in for their technical advice and assistance. It appears that Henry, sensing the importance and complexity of the work at Boulogne, was not yet ready to put full trust in Rogers's direction of the works, but preferred to surround the surveyor with advisers who could prevent him from making serious blunders. As the works proceeded, however, Henry's confidence in Rogers's judgement increased until, as we shall see, he eventually made his surveyor entirely responsible for all of the work in the

[1] Ibid. 424. [2] For a list of the offices and their incumbents, see ibid. 337 (2).

Boulonnais, with express directions to other military and civil officers not to interfere.

The history of the English fortification works at Boulogne may here be retold, not only because it constitutes the most important chapter in Rogers's career as a military engineer, but also because these works played a major role in an important, if relatively brief, episode in Tudor political and military history. For six years the fortification and defence of Boulogne absorbed great portions of the time and the energy, and particularly the money, of the English Government, as anyone may easily discover by perusing the state papers for the period 1544 to 1550. The political side of this episode is sufficiently well-known; but our present concern with the surveyor of the works at Boulogne gives us the opportunity to retrace the history of the actual projects which received so much attention from the king and his ministers, and which devoured so much of the royal treasury. To the political historian who is aware of the deep involvement of the English at Boulogne it should prove interesting to find out specifically what the Government was trying to do in the matter of fortification; only in this way can he understand why it cost the Government the astounding figure of £122,696 for construction work alone (not counting the expenses of the garrison) during the six years that the English occupied Boulogne.[1] To others the study of the works at Boulogne should be interesting because it is possible to reconstruct the appearance of the English fortifications there, in spite of the fact that the structures themselves have long since disappeared.[2] The value of this exercise in 'armchair archaeology' will be further enhanced when we attempt to show, later in this study, that these fortifications constituted an important link in the development of sixteenth-century English military architecture.

A quite detailed history of the works at Boulogne can be written, and the appearance of the fortifications reconstructed, because of the variety and abundance of the source material. Of major importance is the official

[1] This figure comes from a contemporary summary of the expenses of the French and Scottish wars in the reigns of Henry VIII and Edward VI: P.R.O., S.P., 10/15, no. 11, f. 20. I am indebted to Mr. Howard Colvin for calling my attention to this document.

[2] The groundwork for this task has already been laid in an excellent study by Pierre Héliot, 'Les fortifications de Boulogne sous l'occupation anglaise (1544–1550)', *Revue du nord*, xl (1958), 5–38, to which students of the subject will be everlastingly indebted. Although I have worked through all of the original documents, from my own point of view—that of a biographer of John Rogers—I have found little occasion to disagree with M. Héliot's perceptive article.

correspondence preserved in the English state papers; it provides detailed information on the progress of the work and is particularly useful for establishing the dates at which a project was begun and completed. From the French side come several contemporary accounts, in the form either of memoirs by participants in the Boulogne episode, or of 'espials' by French agents seeking information on the progress of the English fortifications. Perhaps most important for understanding the character and appearance of the fortifications are the numerous contemporary drawings which illustrate the works completed or planned by both the English and the French engineers. Practically all of these drawings are undated and unsigned. As we shall see, several of them can be confidently attributed to John Rogers on the basis of documentary, stylistic, and handwriting evidence. As for dating the drawings, they can easily be arranged in chronological order by comparing the successive state of the works as shown on each of them. Providing an absolute date is rather more difficult, but it can be achieved by comparing the drawings with the dates of the works as established through documentary evidence. The arguments for the absolute dates of the drawings will have to be postponed until we have sifted this documentary evidence, but in the meantime we may utilize the drawings for the views which they provide of the works mentioned or described in the English and French accounts. Since most of these drawings are untitled, it will be well to provide a nomenclature for them, in order to avoid circumlocutions. Of the drawings which are reproduced in this study, we shall speak of Plats A, B, C, and D of Ambleteuse (Pls. 15–18), Surveys I, II, III, and IV of Boulogne (Pls. 19–22), and the plat entitled the 'Country of Guynes and Bolenois' (Pl. 14).

Before entering upon a detailed analysis of the individual works, it might be well to review in general the problems which the English faced in planning suitable fortifications for Boulogne. One of these was how to guarantee protection to the ships entering and leaving the haven. Should this lifeline to England be severed, the garrison could quickly be starved out, for, at least in the early days of the occupation, communications overland to Calais were too insecure to guarantee the supply of men and materials from that point. The entrance to the harbour had to be kept open, and, as the English were to discover, this involved combating not only the French, but also natural forces of river and tide which were detrimental to the harbour. Thus Rogers and his fellow technicians found themselves involved in civil engineering projects within the harbour as well as military engineering projects beside the harbour. A second set of problems was connected with the defence of Boulogne itself, the fortifications of which had to be altered considerably in the lower town before the garrison there could readily defend itself. Thirdly, Boulogne was an English enclave in enemy territory, and because of its isolated position, it was advisable to protect the flanks of the English position as far out as possible. Particularly was this true on the eastern and northern sides of Boulogne, for the sea and the river provided at least some protection on the west and the south. Furthermore, series of outlying forts on the east and north would not only deter enemy attacks from those directions, but would also strengthen the lines of communication with the Calais Pale in the north.

All of these considerations were not, of course, immediately apparent to Henry and Rogers and the other Boulogne officials in the autumn of 1544. Only gradually did they recognize all of the problems involved in the proper defence of the newly acquired territory. Our general impression is that the works at Boulogne proceeded largely upon an *ad hoc* basis, with additional works being devised only after some particular problem or crisis had revealed the need for them. More often than not, it was the threat of a French attack from some new direction, or the construction of a new French fortification, that called forth a new project on the part of the English. Indeed, it may be said that the continuing presence of the French in the Boulonnais constituted one of the more important problems facing the English engineers, for this meant that attacks and bombardment from the French had to be endured while the fortifications were actually under construction. This danger was further aggravated at first by the relatively small size of the garrison at Boulogne. Henry removed a portion of his army when he returned to England on 30 September. Three days later the Dukes of Norfolk and Suffolk marched the major part of the English forces to Calais, leaving a garrison of only 3,300 to defend Boulogne. Henry was furious when he heard of their manœuvre, which, he said, was contrary to his express command.[1] Although he ordered the English commanders to return to Boulogne immediately, they dallied in Calais, with their forces daily dwindling through sickness and desertion. As a consequence, when the French executed their famous surprise night

[1] *L. & P.* xix. ii. 347, 352.

attack, or 'Camisade', at Boulogne on 9 October, the English came very close to losing their new possession.[1] The surprise attack on the lower town completely routed the English garrison stationed there. The situation was saved only by the countercharge of the forces from the high town just at the moment when the French commanders had lost control of their own men, who had turned to looting the lower town in the dark of night.

After this disastrous failure, the French army under the Dauphin retired to Montreuil, thereby leaving the English momentarily free to hurry forward the defences of Boulogne. But at the end of January 1545 a large French army under the command of M. Oudart du Bies, Marshal of France and acting lieutenant of Picardy, settled beside the Liane River across from Boulogne. Du Bies's intention was to build a fort facing Boulogne, but the English were successful in driving away his army and forestalling the attempt. Nevertheless, in June the French returned to the same spot and began to construct a large pentagonal fortress. This time the English were unable to prevent the progress of the work. For the next four years the occupying forces on either side of the Liane busied themselves with fortifying to the hilt their respective territories. This was to become a reciprocal process: additional fortifications begun on one side called for new works on the other, and these latter, in turn, inspired additional works to keep the defensive power in balance. Thus, in spite of the fact that we are primarily concerned with the English fortifications and with Rogers's part in their construction, some attention will have to be given to the French works, in order to understand when, why, and how the English works were undertaken.

The necessity of providing adequate protection for the haven early impressed itself upon Henry VIII, even before he left Boulogne in late September 1544. It is surprising, therefore, that he did not immediately see the value of securing both banks of the Liane River at its mouth, for only in this way could he prevent French-established shore batteries from bombarding English ships as they entered and left the haven. (See Pl. 19.) This failure to occupy the southern bank of the river was a costly error, for the French eventually took advantage of the mistake by occupying that shore and building fortifications there. By 5 October Henry had recognized his blunder, and at that time proceeded to reprimand Norfolk and

[1] For the most detailed contemporary account, see the *Commentaires de Blaise de Monluc, Maréchal de France*, ed. Paul Courteault (Paris, 1911), i. 291–300.

Suffolk for having failed to establish an English stronghold 'upon the top of the hill of the other side of the water'.[1] Even then it was not too late to rectify the error, for the French themselves were dilatory in taking and strengthening a position across the river from Boulogne. It appears that the English commanders simply were not anxious to place a contingent of their troops on the other side of the river, where there was as yet no defensive system, and where it would be difficult to send aid quickly in case of a sudden attack by the French. Nevertheless, had they done so the history of the occupation of Boulogne, and of the fortifications built by the English, would probably have been altogether different.

In spite of this failure to take both sides of the Liane, the English still retained an advantage over the French in that the cliff constituting the north bank at the mouth of the river was considerably higher than the south bank, thus giving the English a commanding view over the river and opposite shore. Henry himself perceived this advantage. By early October he was insisting that his engineers fortify this cliff, which was surmounted by an ancient Roman lighthouse known as 'La Tour de l'Ordre', but which the English referred to as 'The Old Man'.[2] Henry wanted a bastillion constructed there before winter. Norfolk reported to the king that his technical advisers—Rogers, Lee, and Cavendish—felt that a suitable bastillion could not be built in such a short time.[3] Henry fired back: 'Though Lee and Rogers might say that it were hard in a short time to make a bastillion to withstand an army, their knowledge had been learned from the King, and the doing of it should not have been relinquished until he had spoken.'[4] Obviously Henry had his own ideas about what could and could not be done at Boulogne. On 26 October he sent to Calais for John Burgate and Thomas Palmer and entrusted them with fortifying the Old Man. Burgate, the surveyor at Calais, was to set out the work, and Palmer was to have 'chief charge and oversight of the same'.[5] Perhaps Henry was still annoyed with Rogers about his opinion regarding the bastillion, or Rogers may have simply been too busy with other works in and around Boulogne to give adequate attention to this important project. At any rate, work got under way at the Old Man, and in an exchange of ideas through letters and plats, the king and his building officials

[1] *L. & P.* xix. ii. 352.

[2] On the origin of this curious English phrase in the Celtic expression *alt maen*, see V. J. Vaillant, 'La Tour d'Ordre et son nom gaulois', published in his *Notes boulonnaises, variétés* (Boulogne-sur-Mer, 1889), 93–113.

[3] *L. & P.* xix. ii. 353.

[4] Ibid. 383.

[5] Ibid. 536.

hammered out the form which the fortification should take. Because of the pressure of time, a temporary bastillion for mounting the cannons and protecting the soldiers was built of timber, stakes, and rods bound together. By the end of November it was reported that within ten days this structure would be ready for the placement of the artillery.[1]

The character of this bastillion can be discerned in Survey I of Boulogne (Pl. 19) which shows a simple, circular parapet around the Tour de l'Ordre.[2] This was the first of what were to be three stages in the development of the fortifications at the Old Man. By the spring of 1545 the second stage was under way, in which a more elaborate scheme of defence was provided for the site. Surveys II–IV of Boulogne (Pls. 20–22) indicate quite clearly the form of this work. Three bulwarks on the landward side of the Tour de l'Ordre were connected by intervening curtains, and the work was surrounded by a ditch. From a description of the English fortifications written by a French spy in 1549 we learn that these bulwarks and curtains contained nothing of stone, brick, or wood, but were made entirely of earth.[3] This perhaps accounts for the fact that the work of repair was more or less continuous at the Old Man throughout the English occupation, for the earthwork suffered destruction not only from French bombardment, but also from the natural elements of rain and winter freeze. Indeed, portions of the fortifications were in danger of falling into the sea, as the French spy pointed out in 1549, for the cliff upon which the fort was placed was gradually being eroded.[4] It is interesting to note that this process, which was ultimately in 1644 to bring the Tour de l'Ordre crashing down, is documented in the contemporary drawings. Surveys I, II, and III of Boulogne show three successive stages of this erosion, as the edge of the cliff slowly encroached upon the Tour de l'Ordre and the circular parapet which the English had built in the autumn of 1544. In fact, the English seem to have played a part in speeding up this process

[1] Ibid. 592, 683.

[2] I have provided in Fig. 3 a schematic plan of the English and French fortifications in and around Boulogne; this may be consulted if the reader has difficulty locating on the various plats fortifications mentioned in the text.

[3] This document, which is entitled 'L'advis de la qualité de ceulx qui sont dedans Boullongne', and which constitutes one of our most important sources for the history of the English works at Boulogne, has been published by Abbé Haigneré in 'Boulogne sous l'occupation anglaise en 1549', *B.S.A.A.B.* i (1864–72), 432–45. In an article with the same name as the document itself, V. J. Vaillant has provided a helpful commentary on some of the obscure points of the document; see *B.S.A.A.B.* vii (1904–7), 334–55.

[4] Haigneré, op. cit. 440.

through their removal of large rocks from the cliff for use in other works at Boulogne.[1]

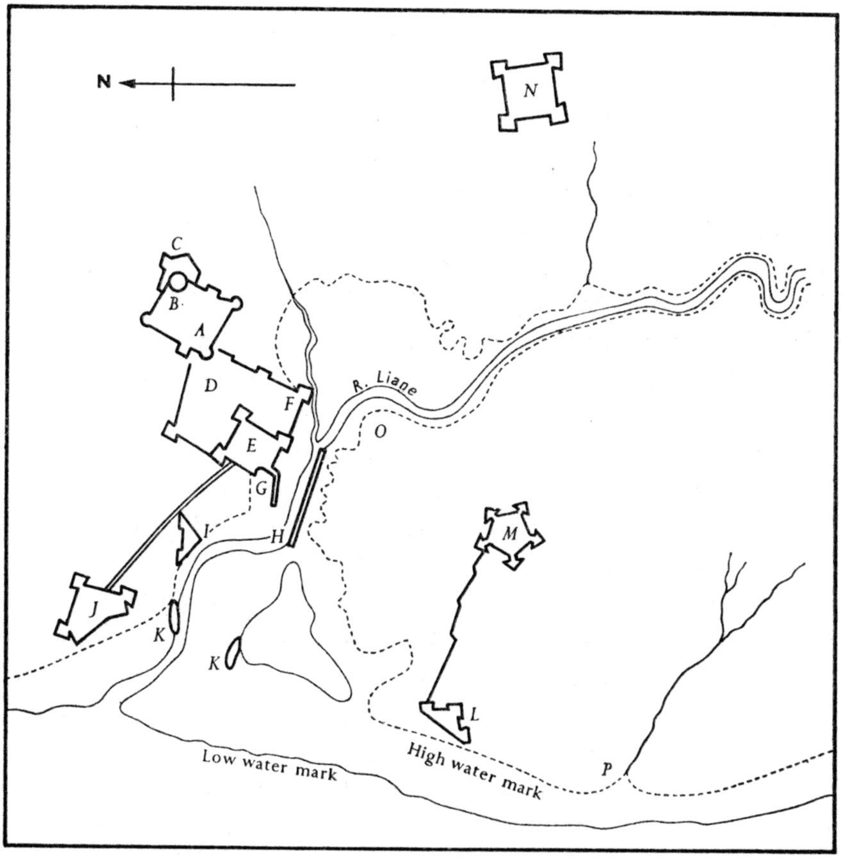

FIG. 3. *Schematic Plan of Fortifications at Boulogne*

A.	High Boulogne	*I.*	Young Man
B.	Castle	*J.*	Old Man
C.	Green Bulwark	*K.*	Jetties
D.	Basse Boulogne	*L.*	Fort Châtillon
E.	Citadel	*M.*	Fort d'Outreau
F.	Clement's Bulwark	*N.*	Boulogneberg
G.	Pier	*O.*	Outreau
H.	Dunette	*P.*	Le Portel (Pawlet)

Considering the military significance which the English placed upon the Old Man, we may be somewhat surprised that this second stage of the fort consisted entirely of earthworks. Several of the other English works at

[1] Haigneré, op. cit. 437.

Boulogne were also built entirely of earth, but the more important forts were encased in brick or stone. The recognition of this weakness resulted in the third stage of the fortification of the Old Man, which began in 1547 and continued intermittently until the end of the English occupation. The direction of the work at the Old Man was once again placed in the hands of John Rogers, when in the spring of 1547 he was charged with repairing and heightening the bulwarks and with building a new gate and bridge facing Boulogne.[1] Apparently this work on the gate, if ever undertaken, proved inadequate, for the next year the earthen rampart at the gate had sunk so low that the point was particularly exposed to French bombardment. Rogers was therefore commissioned to make a gate of stone in the curtain facing towards Boulogne, where the gate would be least endangered by French gunfire. At the same time reference was made to the fact that French shot had pierced the wall of the castle in the Old Man.[2] What was this castle? The French spy in 1549 reported that there had been placed around the Tour de l'Ordre a small fort with four bulwarks, all of them built of brick, with connecting curtains nine feet thick. Surrounding the bulwarks and curtains was a ditch twenty feet wide and twelve feet deep. The spy explained that the English had built this small fort so that if the outer works of the Old Man were captured, the English garrison could retreat into the inner fort, from which they could fire upon the outer works in such fashion that the enemy would be unable to remain there.[3]

Unfortunately, none of the contemporary English surveys of Boulogne show this inner fort. However, a map of the Pas de Calais, drawn in 1558 by Nicholas de Nicolay and published in Abram Ortelius's *Theatrum Orbis Terrarum* of 1570, shows a quadrangular fort—with bastions at each corner—surrounding the 'Tour de Ordre' (Pl. 23). Although the fortresses on this map in general bear at least a rough resemblance to their actual plans, some of them are fairly stylized, and this may be the case here. It is at least certain that the castle was constructed in brick, for remains of two bulwarks were still upon the cliff as late as 1930, when they disappeared with a further collapse of the cliff.[4]

If the fortification of the Old Man was one of the key factors in the protection of the haven, the proper fortification of Boulogne itself was the

[1] *Cal. S.P. For., 1547–53*, 329–30, 335.

[2] Ibid. 352–3.

[3] Haigneré, op. cit. 440–1.

[4] Héliot, op. cit. 32; cf. Haigneré, op. cit. 440, n. 2.

heart of the English defensive system in the Boulonnais. Here the main problems centred upon the different degrees to which the French had previously fortified the upper and lower towns. High Boulogne was a hill-top fortress city which had been given its basic defensive system in the first half of the thirteenth century, when the castle and town walls were constructed by Philip, Count of Boulogne.[1] Around 1415 the exigencies of the Hundred Years War called for a strengthening of the town walls, particularly at the south-west corner, where a bastion was added (La Tour Gayette). A hundred years later bastions were completed at the north-west corner (La Tour Notre-Dame) and the south-west corner (La Tour Française). During the first half of the sixteenth century the four main gates of the town—Porte Gayolle on the east, Porte des Degrés on the south, Porte des Dunes on the west, and Porte de Calais on the north—were strengthened with bulwarks. Then just before the English occupation Francis I constructed a fausse-braye around the outer portion of the castle on the north-east corner of the town.[2]

These fortifications had taken a rather severe battering during the siege of 1544, so that at least one of the tasks facing the English engineers was that of restoring High Boulogne to its previous strength.[3] This work began soon after the English occupied Boulogne, but it appears to have proceeded rather slowly, for as late as April 1546 the breach which had been made in the castle during the siege was still under repair.[4] But the English were not content merely to restore the castle and walls of High Boulogne, for they wished to make their own additions to the fortifications of the town. These were primarily earthworks, and were not such that they radically altered the previous defensive system. The most important work was executed at the north-east corner of the upper town, which had provided the vulnerable point in the siege of 1544. Although the heavily constructed castle was located at this corner, it had proved inadequate

[1] The work can be accurately dated to the year 1231, thanks to a contemporary inscription which provides the date and name of the builder. See Camille Enlart, 'Boulogne monumental', *Congrès archéologique* (Arras, 1880), 301–2.

[2] On the history of these French works, see Pierre Héliot, *Le Château de Boulogne-sur-Mer* (Paris, 1933), 15–18, 32–46; and Camille Enlart, 'Les monuments anciens de Boulogne', *Boulogne-sur-Mer et la région boulonnaise* (Boulogne-sur-Mer, 1899), i. 239–75.

[3] Edward Hall, *The Union of the Two Noble and Illustrate Families of Lancastre & York* (London, 1548), f. 258v, noted that Boulogne was 'so beseged with suche a boundaunce of greate ordinaunce that never was there a more valiaunter assaute made, for besyde the undermyning of the castel, tower and walles, the toune was so beaten withe ordinaunce that there was not left one house whole therein . . . '.

[4] *L. & P.* xxi. i. 565.

because the approach to the walls here was on fairly level ground, thereby providing relatively easy access to the English pioneers who had mined and breached the wall. To offset this condition Rogers laid out a large earthen bulwark that surrounded the north-east corner of the town and extended some distance along the walls towards the Porte de Calais and the Porte Gayolle. Called the 'Green Bulwark' by the English, it can be seen on Surveys II–IV of Boulogne. That the bulwark shown on these three drawings was an English addition can be demonstrated by comparing the drawing with Survey I, which apparently was drafted before the Green Bulwark was constructed. Additional information on the Green Bulwark is provided by the French espial of 1549, where it is noted that the bulwark was built of earth and faggots and was sixty feet 'large'—apparently meaning that it was sixty feet from the castle or town walls to the outer edge of the bulwark. Surrounding the bulwark was a ditch twenty-four feet wide which contained water to a depth of three feet.[1]

From the same source we learn something about the other earthworks constructed by the English around High Boulogne. The ditches and fausses-braies surrounding the town were left much as they had been constructed by the French, but outside these the English built several 'platforms' of earth and faggots.[2] One of these was constructed near the Porte Gayolle; another stretched along the southern wall of the town, from the Tour Française around to the Porte des Dunes on the western wall. Outside this latter gate was built another platform, and yet another was placed between the gate and the Tour Notre-Dame on the north-west corner of the town. Although the one between the Tour Française and the Porte des Dunes is reported to have been sixteen feet high and two hundred feet long, these mounds of earth and faggots seem to have been of rather simple design and construction. Their chief purpose was to provide placements for artillery outside the medieval town walls, which were probably unsuitable as mounts for the English cannons. The wisdom of this plan is illustrated by the fact that, upon the recovery of their town, the French developed the English 'platforms' into a fully articulated system of earthen outworks.[3]

[1] Haigneré, op. cit. 434. This ditch is clearly shown on Surveys II and III.

[2] Ibid. 434–5; 'ilz ont faict une platte forme de terre et de fagotz', remarked the French spy.

[3] The later system is shown on a plan of High Boulogne dating from 1677 (Paris, B.N., Est. I^d. 17), reproduced in A. de Rosny, *Album historique du Boulonnais* (Neuville-sous-Montreuil, 1892), Pl. XXIV. Héliot at first thought (*Château de Boulogne*, 19, 37) that these more elaborate works—remnants of which are still visible along the Boulevards Mariette and

But it appears from the description of the French spy that even the English earthworks were so numerous and extensive that an enemy force approaching High Boulogne from almost any direction would find itself looking into the muzzles of English cannon.

Basse Boulogne presented to Rogers and his fellow engineers considerably greater problems in providing adequate defence, for the walls of the lower town were inferior to begin with and had been badly damaged in the siege. But the town played a major part in the English defensive system, for it bordered upon the Liane and was the point where the English unloaded supplies from their ships entering the haven. In order to protect both the town and the ships some more elaborate fortification was needed than the inadequate earthen ramparts inherited from the French. By July of 1545 this work had been at least partly decided upon, and Rogers was busily directing the construction of a stone bulwark in the south-west corner of the town, which the English henceforth referred to as the Citadel.[1] Although this fort is absent from Survey I, it figures prominently on Surveys II and III, from which we can get a clear notion of its form. Quadrangular in shape, the Citadel had bastions at three corners, while a 'pere' or dock provided a protected unloading place at the fourth corner. The significance which the English attached to the Citadel is indicated by the fact that it was built in stone, rather than in the earthwork which had been characteristic of English building in Boulogne up to this time. It was decided to make the structure of stone not only because of the important part it played in protecting the ships during unloading, but also because the Citadel was one of the main English arsenals in the Boulonnais. Furthermore, it was particularly subject to bombardment from 'Fort d'Outreau', which the French were building directly across the haven from Basse Boulogne. Indeed, French guns were pelting the Citadel even while it was under construction, but it was reported on 17 July that 'now Englishmen have learnt to contemn the gun shot'.[2] This stoical attitude arose from English confidence that the French gunners were incapable of hitting anything at which they aimed, because of the great distance from

Prince Albert—had been constructed by the English, but then he concluded ('Fortifications de Boulogne', 22) that these were later French works. Surveys II and III of Boulogne support Héliot's conclusion, for the plans of High Boulogne on these drawings do not show the bastioned outworks of the drawing of 1677, but suggest, if anything, that the English platforms were merely simple, rectilinear mounds of earth.

[1] *L. & P.* xx. i. 1160.

[2] Ibid. 1210.

Fort d'Outreau to the mouth of the haven and to Basse Boulogne on the other side of the river. On a visit to Boulogne in December 1545 one of Henry's secretaries of state, William Paget, reported sarcastically that 'the Frenchmen's new fort shows proudly, but is so distant that there a white horse can hardly be discerned from a black, and their shot is only troublesome for the noise'.[1] He was exaggerating just a bit; nevertheless the fort had not proved to be the great threat which it seemed to English officials back in July.[2]

Still the relative ineffectiveness of Fort d'Outreau did not prevent the English from further fortifying the lower town, for of course they had not only to fear French cannonading, but also the possibility of another French attack like that of the Camisade. The town was particularly vulnerable because of its inadequate ramparts. This was recognized by a commission consisting of the Earl of Hertford, John Rogers, Richard Lee, and two others, appointed in April of 1546 to inspect the fortifications at Boulogne and to make recommendations for further works. The commission suggested that either the lower town should be razed and abandoned or else better ramparts provided, for, indefensible as it was at the moment, the town constituted a danger to the English position.[3] With the alternatives now more clearly defined, it was decided to fortify Basse Boulogne rather than abandon it. The Citadel provided an anchor on the south-west corner; the south-east corner was anchored by another stone bulwark, which had already been started by the end of March 1546.[4] This latter project, known as Clement's Bulwark, was rather slow in building, for by 1549 it had not yet been completed.[5] Its form, however, can be ascertained on Surveys II–IV. In order to complete the defences of Basse Boulogne, Rogers and his corps of pioneers erected ramparts around the town which connected the Citadel with Clement's Bulwark and with High Boulogne.

[1] *L. & P.* xx. ii. 919.

[2] J. Hector de Rosny, *Histoire du Boulonnais* (Amiens, 1871), iii. 205, castigated the French commander, Oudart du Bies, and the Italian Engineer, Antonio Melloni, for their poor judgement in planning the site of the fortress. They did commit a strategic blunder by not placing the fort at the mouth of the haven, but it is doubtful that they could have located Fort d'Outreau closer to the river and thereby to Basse Boulogne, because of the topography of the south bank of the river. Cf. Surveys II and III of Boulogne, where the scale, 'the ynche stondethe for v^{c} ffote', makes it possible to determine that Fort d'Outreau was 3,500 to 4,000 feet from the Citadel. This was certainly not beyond the range of sixteenth-century cannons, although the accuracy at this distance cannot have been very great. On artillery range in this period, see Michael Lewis, *Armada Guns* (London, 1961), 37–39.

[3] *L. & P.* xxi. i. 565.

[4] Ibid. 481.

[5] As reported by the French spy: Haigneré op. cit. 436.

The French spy reported in 1549 that these walls were earthworks, outside which was a ditch some thirty feet wide.[1] This certainly must have been the case for those ramparts extending from the stone bulwarks to High Boulogne; but a careful study of Survey III suggests that the rampart lying between the Citadel and Clement's Bulwark was made of stone. Since this section of the town walls would receive the severest battering from Fort d'Outreau, it would have been more natural to face this rampart with stone, rather than to make it entirely of earth.[2]

By the spring of 1546 the English defensive system at Boulogne was beginning to take shape. The Old Man was the protective arm at the entrance to the haven; High Boulogne provided the headquarters of the English garrisons and the heart of the system; Basse Boulogne was the stomach, that received the supplies from English ships which sailed into the mouth of the Liane River. Although the lower town had acquired its own armour in the form of the Citadel and Clement's Bulwark, still the English felt that it needed a protective shield, which was provided in the nearby fortress known as the Young Man. This latter work was devised in the earlier stages of the fortification of Boulogne. It is mentioned in a letter of 17 July 1545,[3] and its placement (though not its ultimate form) is shown on the earliest English survey of Boulogne, where it is to be found about half-way between the Old Man and Basse Boulogne (Pl. 19). Yet it suffered an unhappy history of stop-and-go work. In December 1545 William Paget reported that 'the Young Man (if he were finished) liketh my foolish judgment well'.[4] Only four months later the previously mentioned commission of Hertford, Rogers, Lee, *et al.* recommended that the Young Man should be thrown down, since it would take a long time in building and would require a large garrison when completed, and yet 'could do no more than a square tower of 30 ft. broad kept by 10 men'.[5] In August of 1546 the fate of the fortress had not yet been decided, but in September the answer came that it 'shall go forth out of hand, as the

[1] Haigneré, op. cit. 437.

[2] The details of Survey III, by the way, indicate that the walls of the Citadel within Basse Boulogne were constructed in earth, without stone revetments. This view is contradicted by the French spy, who remarked that the Citadel 'est tout édiffié de pierre' (ibid. 436). But the fact that at the restitution of Boulogne in 1550 the Citadel was so easily dismantled when the property was restored to the Franciscan convent [Vaillant, 'L'advis de la qualité de ceulx qui sont dedans Boullongne (1559)', *B.S.A.A.B.* vii (1904–7), 351–2] indicates that the spy may have been mistaken about the character of the walls of the Citadel which were inside the town walls.

[3] *L. & P.* xx. i. 1210.

[4] Ibid. xx. ii. 919.

[5] Ibid. xxi. i. 565.

surveyor of the works can declare'.[1] On 1 December Edward Dudley, one of the councilmen at Boulogne, optimistically remarked that 'the building of the Young Man is now in hand and shortly I shall be appointed to lie there'.[2] Yet nearly three months elapsed before Rogers was able to get the work under way. On 21 February 1547 Rogers himself wrote a letter to William Paget, explaining what he was doing at the Young Man:

> After my duty considered, it may please your mastership to understand that within the Young Man I have traced or staked forth a little castle and, as I think, very fit for the same place if it may please the king's majesty's Council that it may be gone in hand withal, as Mr. Dudley is not a little fain thereof; but till I know your pleasures it shall stay and nothing be done thereto. A plat thereof have I sent unto you by this bearer, every inch containing ten foot. The one point standeth directly to the hill where the windmill standeth, the other point to the French fort. The gate openeth right towards Basse Boulogne. . . . And then no doubt though the plat seem tedious to do, yet will it lightly be made and no great charge—neither shall it let any other work.[3]

The work which Rogers outlined in this letter continued, but at a dawdling rate, for in May of 1548 the fortress was so far from completion that the ordnance could not be placed in it, since the platforms for the guns had not yet been erected.[4] Nevertheless, the Young Man was completed by the time the French spy made his report in 1549, for he says that the platforms were up and were armed with four pieces of artillery. He adds rather contemptuously that the Young Man was built with brick, that its walls were only 6 feet thick, and that it 'n'est pas pour résister à coups de canon'.[5] The purpose of the fort, remarks the spy, was to give support to the Old Man and to prevent the French from gaining a foothold between that point and High Boulogne. Furthermore, its location overlooked Basse Boulogne and the haven in such a way that it afforded protection for both: no enemy ship could remain in the haven with the Young Man in English hands.

Although the purpose and placement of the Young Man is known, the exact form of the completed fort constitutes an elusive problem. Survey I

[1] Ibid. 1487 and xxi. ii. 54.

[2] Ibid. xxi. ii. 482.

[3] P.R.O., S.P. 68/13, no. 42. (*Cal. S.P. For., 1547–53*, 306.)

[4] *Cal. S.P. For., 1547–53*, 346.

[5] Haigneré, op. cit. 438. The Young Man is referred to in this document as the 'fort de Paradis', and in the French tradition at Boulogne as 'Le Fort Rouge'. The fort has since entirely disappeared, although some of its foundations were discovered in the early part of the nineteenth century (ibid. 438, n. 4).

of Boulogne shows it as a rectangular 'medieval' castle in much the same form as High Boulogne. Survey II represents it in triangular form with bastions at two of the angles, and much the same form is shown on the plat of the 'Country of Guynes and Bolenois' (Pl. 14). Curiously enough, Survey III of Boulogne, which is the most detailed of all these drawings, does not show the Young Man at all.[1] Unfortunately, the detailed plan of the Young Man which Rogers drew to a scale of one inch to ten feet, and which he mentioned in the above-quoted letter, does not appear to have survived. It is probably safe to assume, however, that the Young Man was a triangular fortress, and thus would have been related to the Old Man in form as well as name.

As the various works in Boulogne proceeded in the spring of 1545, it was realized that some effort should be made to guarantee communications between these strongholds. Accordingly, Henry VIII recommended that a deep ditch be cast down the hill from the Old Man and a bulwark of earth built up towards High Boulogne. This protective work would help defend Basse Boulogne from the landward side to the north-west, and would allow safe passage of Englishmen between the lower town and the Old Man. Henry asked for Lord Poynings's opinion on the matter.[2] The basic idea proved to be acceptable, but when it was actually executed the ditch ran from the Old Man directly to the Citadel, as can be seen on Survey II. It would appear from Survey III that this simple ditch was later replaced with an elaborate arrangement of ditch and bastioned ramparts, but there is good reason for doubting that this drawing represents in this case an earthwork actually constructed.[3]

Because Fort d'Outreau was only the width of the river away from Basse Boulogne, the English found it necessary not only to fortify the lower town and the Young Man, but also to build works in the harbour itself for the protection of their ships. This need was recognized as early as September 1545 when the Privy Council sent a resolution to the officials at Boulogne, suggesting that a jetty or mole some 400 feet in length be built into the haven, 'to defend the victuallers from shot of the enemy's new fortress'.[4] By April of 1546 this project appears to have been nearly completed.[5] The French spy in 1549 described it as a wall 'entering the haven

[1] This peculiarity can be explained in connexion with the dating of the drawing. See below, p. 102.

[2] *L. & P.* xx. i. 325.

[3] See below, p. 103.

[4] *L. & P.* xx. ii. 323.

[5] Ibid. xxi. i. 565.

—three hundred feet long, sixteen feet wide, and twenty-six feet high—which covers and defends all of the ships entering the port in such a way that the French gunfire is unable to harm them'.[1] Surveys II and III of Boulogne indicate quite clearly the nature of this 'pere', as it is called on the latter drawing. It was built of stone and constructed so that at high tide English ships could snuggle into the corner between the 'pere' and the Citadel and be safely unloaded.

Besides protecting English ships from enemy gunfire, this mole also prevented sand from silting up this part of the haven. The protection of the harbour from natural forces of river and tide was one of the continuing problems which the English faced at Boulogne. To meet this problem, two jetties had been built at the mouth of the haven—one of wooden piles, the other of stone. In June of 1546 it was noted that the jetty of piles, 'although on the other side of the water, belongs to the King, and if it were maliciously marred a storm might fill the haven with sand'.[2] Both jetties are shown on Surveys II and III, whereas Survey I indicates only the stone jetty on the English side of the haven. Since this latter drawing shows only the earliest of the English fortifications, it suggests that the stone jetty had either been inherited from the French or had been one of the first works undertaken by the English.

No doubt the most important English project within the haven for protecting it from both natural forces and French cannonading was the jetty which came to be called the Dunette. This work was built of stones taken from the cliff at the Tour de l'Ordre.[3] Although the English began to collect these stones in January of 1547[4] it was not until 1 April that the first stone was laid, as we learn from a remark in one of John Rogers's letters.[5] The Treaty of Camp, concluded in June of 1546, had forbidden new fortification works to be started by either the French or the English. When work began on the Dunette the French immediately became suspicious that the English were subverting this clause in the treaty. In May Lord Grey assured Gaspard de Châtillon, the French commander of Fort d'Outreau, that the new work was only a jetty and not a fortress. Châtillon replied that this might be true, but the size and character of this 'jetty' would make it easy to transform it into a fortress in time of war.[6]

[1] Haigneré, op. cit. 436.
[2] *L. & P.* xxi. i. 1048.
[3] Haigneré, op. cit. 437.
[4] *L. & P.* xxi. ii. 715.
[5] P.R.O., S.P. 68/14, no. 103. This remark is not calendared in *Cal. S.P. For., 1547–53*, 327.
[6] *Cal. S.P. For., 1547–53*, 337.

Meanwhile in London the Duke of Somerset was trying to convince Odet de Selve, the French ambassador, that the Dunette was a simple jetty with no pretence of being a fortress, and that it was being built in order to make the harbour safe for the entrance of ships, some of which had already been lost as a result of the poor condition of the haven.[1] The French were not satisfied, however, and continued to complain about the work as it progressed, while Somerset persisted in his argument that the Dunette was not a fort. In July of 1548 he added the specious argument that the Dunette was merely a continuation of the work planned by Henry VIII *before* the Treaty of Camp, when the 'petite mole' adjacent to the Citadel had been begun in the winter of 1545/6.[2] Apparently finding this tack ineffective, he returned the next month to his claim that the Dunette was merely a jetty for controlling the force of the tide within the harbour.[3]

By this time it had become obvious to all that the Dunette was rather more than a jetty. The French spy observed it in 1549 and described it as a long and narrow fort which was 20 feet wide and 36 feet high; at each end was a wall from which soldiers could fire upon those who came near the fort.[4] When the French king, Henry II, attacked the Dunette in 1550 he described it in almost lyrical terms: 'It is a great long jetty of large, coarse stones, brought together and walled up with great skill. It contains handsome quarters for the soldiers and is properly constructed on top for placing artillery, of which there are a great many pieces. One may well say that since the time of the Romans there has not been built in this place a sturdier or more superb edifice than this, which they call the Dunette.'[5]

These verbal descriptions of the Dunette can as usual be supplemented with the contemporary English and French drawings relating to Boulogne. Particularly important is the drawing which, from the inscription placed upon it, may be called 'The Trew Conterfaicte of the State of the Pere'.[6] It has generally been assumed that this is a view of the Dunette, and since the specific date of 26 January 1549 is given on the drawing, we probably have here a definite representation of the structure as it appeared

[1] Odet de Selve, *Correspondance politique d'Odet de Selve*, ed. Germain Lefèvre-Pontalis (Paris, 1888), 148.

[2] Ibid. 407.

[3] Ibid. 431.

[4] Haigneré, op. cit. 437.

[5] A. de Rosny, 'Documents inédits ou rarissimes, concernant les sièges de Boulogne, 1544–1549', *M.S.A.A.B.* xxvii (1912), 533.

[6] B.M., Harleian Roll E. E. 14. This drawing is ten feet in length and one foot ten inches in width, and therefore cannot easily be reproduced in one photograph. Since it is not one of Rogers's drawings, I have refrained from reproducing even a portion of it in this study. It is, however, available in a fold-out plate in Rosny's *Album historique*, Pl. XVII.

at this particular time. In general, it may be said that the drawing conforms to the descriptions given by the French spy and Henry II, for the large unhewn stones are seen embedded in the masonry walls, and the lodgings and placement of the artillery are shown on top of the Dunette. In this drawing the Dunette does appear as quite an impressive structure, just as Henry II described it. But when one compares this drawing with the surveys of Boulogne, an interesting and at first puzzling fact emerges, for the 'jetty' shown on the latter drawings has quite a different character from that of the 'Trew Conterfaicte'. Survey I pictures the haven at full tide, while within the haven—and running almost its full length—is a rather nondescript mole. Surveys II–IV show a long jetty opposite Basse Boulogne which appears to consist of two rows of piles driven into the river bed, with a fill between them. Since it is known that the Dunette was begun in the winter and spring of 1547, and since all of these surveys predate this period (as we shall see), it must be concluded that a wooden pile-jetty preceded the stone Dunette. I have found no reference to the building of such a jetty by the English, so that it seems fairly certain that they inherited from the French this one shown on the surveys. But it does seem strange that in their arguments with the French over the Dunette, the English did not make a point of the fact that they were merely replacing a wooden jetty with a stone one.

If the Old Man was a protective arm reaching out from Boulogne to the edge of the sea, the other arm stretching inland along the river ended at the fort which the English called Bullemberg (Boulogneberg), and which was located on Mont Lambert, a few miles east of Boulogne.[1] Ever since the siege of Boulogne the English had maintained on this hill overlooking the river a small garrison which they referred to as the Master of the Horse's Camp. But they made no very serious efforts at fortifying the camp until the late spring of 1546, when the French began massing forces across the river at St. Étienne. It was feared that from this point south-east of Boulogne the French would cross the Liane over the Pont de Brique and advance along the unprotected north bank of the river (see map on Pl. 23). Rather than wait for this eventuality, William Lord Grey, governor of Boulogne, took the offensive by marching his own forces across the Pont de Brique and skirmishing with the French near St. Étienne.[2] Meanwhile,

[1] The identification of Boulogneberg with Mont Lambert is demonstrated by a reference in Odet de Selve, op. cit. 433.

[2] *L. & P.* xxi. i. 892.

Grey urged Rogers to hurry forward the fortification of the Master of the Horse's Camp, which he felt was a key position in the holding of Boulogne. Not only would it protect the north bank of the Liane, it would also guarantee access to the woods and pastures east of Boulogne which the English sorely needed to supplement their supplies coming in by ship.[1] But as usual, the initiation of fortification works on one side produced reciprocal reactions on the other. Seeing the work proceeding at Boulogneberg, the French army began not only to occupy but to fortify St. Étienne. Or was it the other way around? Both sides claimed that their works were precipitated by the activities of the other. As we shall see, the issue became particularly important when the Treaty of Camp, signed in June 1546, expressly forbade the construction of new fortifications on either side.

During the summer of 1546 Rogers pressed forward the works at Boulogneberg with all haste. In early June it was reported that the work was a princely fortification, with ditches 'marvelous deep and wide', and it was thought that within three or four weeks the fort could be made defensible against anything which the French could field.[2] As usual with building construction, the completion date was over-optimistic. By mid August Rogers was being urged to get the thing ready at least for the winter.[3] In January of 1547 the fort still lacked an adequate gate and water supply, nor did it have sufficient storage space for supplies and munitions.[4] The first two points seem to have been fairly quickly met, but in February the inadequate storage space was again noted, and in April Rogers was particularly instructed to correct this matter.[5]

We know from the French espial of 1549 that Boulogneberg was an earthen fort and that it had four sides, similar in form to the Citadel, though somewhat smaller.[6] Curiously enough, none of the contemporary English drawings contains a plan of Boulogneberg, although one French drawing dating from 1548 does show the fort in the upper right-hand corner behind High Boulogne.[7] Unfortunately, the drawing is torn at this corner and part of the fort is missing, so that little can be made of its form. But a good representation of 'Bollemberg' is given on Nicolay's map of the Pas de Calais, where the rectangular form of the fort is clearly shown

[1] *L. & P.* xxi. i. 1016. See also ibid. xxi. ii. 29.

[2] Ibid. xxi. i. 1017.

[3] Ibid. 1472; cf. Dasent, *A.P.C.* i. 519.

[4] *L. & P.* xxi. ii. 731.

[5] *Cal. S.P. For., 1547–53*, 294, 329.

[6] Haigneré, op. cit. 435–6.

[7] Boulogne, Bibl. de Boulogne, Album in plane no. 1, pl. 1 (reproduced in Rosny, *Album historique*, Pl. XVI).

(Pl. 23). From this it can be seen that the French spy's comparison of Boulogneberg with the Citadel was accurate, for at each of the four corners was placed a bastion, much like those of the fortress within Basse Boulogne.[1]

By the summer of 1546 the various works in and around Boulogne had secured the haven in English hands, in spite of the French fortresses that were being built across the River Liane. But the maximum security of Boulogne could not be established merely by keeping the harbour open to English shipping. It was also desirable to confirm overland communications between the newly acquired Boulonnais and the older English establishment in the Pale of Calais. To this end fortifications were proceeding at Ambleteuse and Blackness—both of these being small towns on the coast of the Pas de Calais between Boulogne and Calais. (See Pl. 23.) The two towns had come into English hands by the spring of 1546, although I have been unable to determine exactly how and when this came to pass.[2]

The topography of the area around Ambleteuse is shown in five contemporary plats which were produced in connexion with the English decision to fortify the site. Thanks to this pictorial evidence one can visualize quite clearly the natural conditions of the harbour and its surroundings and thereby the problems which the English engineers faced. But because it can seldom be decided *a priori* whether a sixteenth-century plat represents, on the one hand, a plan for construction which may or may not have been executed, or on the other, a survey of a completed building, care must be exercised in using these drawings as evidence for the works actually carried out by the English at Ambleteuse. Particularly is this true with these plats, since they present four schemes differing more or less from one another. As hitherto in this study, it will be necessary to engage in a careful and detailed analysis of the drawings in the light of the documentary evidence that is available. And as usual, it will be advisable to provide an abbreviated nomenclature for easy reference to the drawings. Plat A of Ambleteuse (Pl. 15) is a survey of the haven which shows no fortifications. Plat B (Pl. 16) is a similar view which includes a

[1] B.M., Cott. MSS., Aug. I, Supp. 13, is a plat of a quadrangular bastioned fort that may represent Boulogneberg, but there are no inscriptions, endorsements, or other signs to guarantee this supposition. The drawing style makes it very unlikely that it was drafted by Rogers.

[2] References to Ambleteuse under English occupation begin in March (*L. & P.* xxi. i. 394) and to Blackness in May (ibid. 960).

fort with three large bulwarks, while Plats C (Pl. 17) and D (Pl. 18) show a fort with five bulwarks. Another plat (not reproduced here) also depicts a pentagonal fort with bulwarks at each corner.[1]

Contemporary letters indicate that a number of plats of Ambleteuse were produced soon after the decision was made to fortify the town; it is likely that some or all of the above-named drawings are among those mentioned in the letters. By 15 March 1546 the English had occupied Ambleteuse.[2] In addition to the fact that a fort at Ambleteuse would secure overland communications between Boulogne and Calais, it was recognized that the town, located at the mouth of a river, might provide another harbour besides that at Boulogne for bringing men and supplies into the Boulonnais. But there were misgivings about the quality of this haven, so that it was necessary to survey the natural conditions of the site to see if it would be usable as a port. Sir Thomas Wyatt and John Rogers were assigned to make this survey, which they completed on 27 March. Their report was not very comforting, since at that moment the tide was 'dead neap' (that is, at its lowest ebb). Consequently the harbour, while 400 feet in length, was only one fathom (5 to 6 feet) deep; in the river before the town there was a mere 4 feet of water.[3] These conditions are set forth in Plat A of Ambleteuse (Pl. 15). Although there is no mention of a plat in the report of Rogers and Wyatt, this drawing so nearly conforms to their report that one suspects it was produced by one of them at the time of the report or shortly thereafter. Its purpose would have been to provide a graphic illustration of the topography of Ambleteuse for those who were unfamiliar with the area but who were nevertheless involved in planning the use of the haven. The fact that no fortifications are shown suggests that the drawing was made at an early date, before any definite building plans had been formulated.

Yet those plans were rapidly coming into being. A few days after Rogers's and Wyatt's survey of the haven at Ambleteuse, an army arrived under the command of Lord Lisle, and discussion ensued regarding the fortification of the town. Lisle brought a plat from Henry VIII which he showed to Hertford, Rogers, Lee, and others. Although they 'all commended it' (as well they might!), there was doubt whether the fort should have five bulwarks, as Henry's plat showed, or four. With due consideration of the king's plan, they nevertheless decided that the ground was

[1] B.M., MS. Facs. 372, vol. i, no. 54. [2] *L. & P.* xxi. i. 394. [3] Ibid. 471, 472.

better suited for four.[1] This decision was forwarded on 1 April by Lisle, who remarked that Hertford doubtless would send a plat to the king showing the conclusions of the experts at Ambleteuse. Apparently Henry was not satisfied with Lisle's explanation or with the plat—if one was sent to the king—for he soon requested Rogers to send a 'plat of the whole'. In a letter from several local officials, dated 7 April, Rogers begged off the request, saying that he was too busy removing old walls and keeping pioneers at work. He suggested that Lee, who would soon be returning to England, would be able to explain the matter personally, so that the king would be fully satisfied with the decision. Besides, he remarked, the king already had a plat which would help to explain the matter.[2]

Plat B of Ambleteuse (Pl. 16) shows a fort with three large corner bulwarks and a much smaller one on the curtain along the edge of the river. If it is permissible to interpret this plan as a fort with four bulwarks, then Plat B may be the drawing mentioned in the letter of 7 April. It would then have been drafted between that date and the preceding 30 March, when Hertford, Rogers, and Lee had decided upon a four-bulwark rather than a five-bulwark plan. In any case, there can be little doubt that this plat was drafted by Rogers, for he has placed his monogram, ℞, in the lower right-hand corner of the drawing. This unusual clue provides proof positive that the drawing was by Rogers. The resulting confidence regarding this plat will make it easier to attribute to Rogers certain drawings which show strong similarities to this one in subject-matter, drafting style, and handwriting, but which lack the dramatic proof that Plat B possesses. For instance, the similarities of Plats A and B regarding the three factors just mentioned make it possible to confirm our earlier conclusion, which was based merely on circumstantial evidence, that Rogers drafted Plat A.

In spite of the explanations given by his officials and the plats produced by Rogers, Henry seems to have insisted that his engineers think in terms of a fort with five bulwarks. At least this is the implication of the fact that there are extant three contemporary plats of Ambleteuse which show a pentagonal fort. Two of these drawings (Plats C and D: Pls. 17 and 18) also display a bulwark on the opposite shore that is connected to the fort by a barrage built across the river. The similarity in the overall design of the fort as seen on these plats is obvious, while it is equally clear that the plats were drawn by different hands. But closer comparison reveals

[1] Ibid. 507. [2] Ibid. 565.

another interesting point—that one of the drawings is to some extent a copy of the other. Otherwise it is inexplicable how two draughtsmen, displaying rather different drawing styles, would have produced a plat of the harbour and fort from exactly the same angle of vision, with roads, river banks, and contour lines of hills and sand dunes falling exactly into the same patterns. So the question occurs as to which of the two drawings came first.

Plat D clearly provides more details regarding the fort, where not only are the internal streets carefully laid out in a formal manner, but also towers are rather curiously placed in the gorges of the bulwarks; furthermore, the flanks of the bulwarks are more complex than those on Plat C. In short, Plat D appears to be a more elaborate plan of the fort than is shown on Plat C. From a logical point of view one might argue that the simpler plan would precede the more complex; *ergo*, Plat D is based on Plat C. In considering the topographical features of the two drawings, one is led to the same conclusion. While both show the same features and follow the same outlines, Plat C includes details and displays precision in the delineation of topographical features that Plat D lacks. Assuming that one plat is based upon observation of these features *in situ*, while the other plat is a mere copying of these observations, it will be seen that once again, logically speaking, Plat C would have preceded Plat D. But apart from these merely logical arguments, there is a letter from Hertford to the king, dated 27 April 1546, which states that in response to the king's request a plat of the fortifications and haven at Ambleteuse had been produced. Hertford adds that the plat was being brought to the king, along with this letter, by John Rogers, who would explain the drawing to the king.[1] Plat C may well have been the drawing in question. The similarity in drafting technique to the other two plats of Ambleteuse produced by Rogers (Plats A and B) strongly suggests that this plat was likewise by him. Furthermore, the drawing is devoid of inscriptions except for two brief ones that were obviously added by later hands. This fact fits nicely into the circumstances indicated in the letter, namely, that Rogers was personally carrying the plat to Henry. Hence explanatory inscriptions would have been unnecessary.

From this analysis of the plats of Ambleteuse and the relevant documentary evidence, it is possible to reconstruct the series of events which

[1] *L. & P.* xxi. i. 686.

led to the production of the drawings. Rogers and Wyatt surveyed the harbour at Ambleteuse on 27 March, and at that time or soon after Rogers produced Plat A as a graphic description of their survey. Meanwhile, Henry had drawn a plan (apparently no longer extant) for a pentagonal fort Ambleteuse, but this was rejected by Rogers, Lee, and other technicians on 30 March. Rogers then drafted Plat B as an alternative to Henry's scheme, but the king persisted in favour of a pentagonal fort. By 27 April Rogers had yielded to the king's view and produced Plat C, which showed not only a pentagonal fort but also an additional bulwark and barrage thrown across the river. This plat was presented to the king by Rogers in person, and then someone (possibly Henry himself) redrafted it in the form of Plat D, which displayed a more elaborate internal arrangement of the fort without changing its basic outline.

Incidentally, this internal arrangement of Plat D is worth noting, because it clearly reflects a current interest in the radial plan for fortified towns. The radial town, first introduced into European town-planning by Italian architects of the fifteenth century, had only recently been revived by Italian military engineers.[1] In 1544 Girolamo Marini began for Francis I the fortress town of Villefranche-sur-Meuse, north-west of Verdun, in which a quadrangular scheme with four large bastions was laid out with streets radiating from a central plaza.[2] The similarity of the street plan to that on Plat D of Ambleteuse is striking, even though the two towns were based on two different geometrical forms—the square and the pentagon. It may very well have been an Italian engineer at Henry's court, or merely someone familiar with Marini's work at Villefranche, who inspired the redrafting of Rogers's plan for Ambleteuse (Plat C) in order to produce the radial plan of Plat D. No doubt Henry would have been enthusiastic for this suggestion that would make his fortress town at Ambleteuse as up to date as the one his arch rival was building at Villefranche-sur-Meuse.

If the dating of Plat C in late April of 1546 is correct, it means that the drawing was made at a very early stage in the construction work at Ambleteuse. As usual, this raises the question of whether or not the fortifications were actually built in the form shown on Plats C and D. Unfortunately, documentary evidence is scanty on the subject; it seems likely that some of the correspondence between Henry and his technical officials and

[1] Horst de la Croix, 'Military Architecture and the Radial City Plan in Sixteenth Century Italy', *Art Bulletin*, xlii (1960), 263.

[2] Ibid. 276 and Fig. 15.

advisers at Ambleteuse has not survived, for there is no indication in the surviving letters as to when or how Rogers and his colleagues became convinced that Henry's proposal for a pentagonal fort was superior to their plans for a fort with four bulwarks. But there can be little doubt that Henry's view ultimately prevailed. An ordnance survey of 1547 lists five 'mounts' or bulwarks of the fort at Ambleteuse, indicating that at that time the fort was pentagonal.[1] Furthermore, a plat of Ambleteuse apparently drawn later than Plats C and D also shows a pentagonal fort—although not in exactly the same position as on the other two plats.[2] But on this later plat there is an additional rampart extending from the fort to 'The New Bulwarke' located in the sand-hills along the seashore.[3] Since there is little documentary evidence for this other rampart and bulwark—the ordnance survey of 1547 does not mention it—one might be inclined to reject out of hand the idea that these structures were actually built by the English. But a plan of Ambleteuse published by Nicholas de Fer a century and a half after the English occupation raises some intriguing questions.[4] De Fer's view shows the pentagonal fort with almost exactly the same outline and in the same location as shown on Plats C and D. In the cartouche on the plan this inscription occurs: 'Ambleteuse Ville ancienne du Boulenois. . . . Quand ce port auquel on trauaille Sera achevé, il Sera un des plus Beaux et des plus Sûrs de l'Europe.' It is clear from this that the French were continuing the task begun by the English of making Ambleteuse into a well-fortified port, and it is interesting that they were following the plan which Rogers, Henry VIII, and others had devised long before. Another point of interest is that de Fer's plan shows a rampart extending from the fort to a bulwark by the sea, and both the rampart and the bulwark are positioned very closely to those shown on the later plat of Ambleteuse. Thus it may be that, after all, these works were built by the English and later utilized or reconstructed on the same lines by the French.

De Fer's late seventeenth-century view of Ambleteuse raises another interesting point, for it shows a well-developed haven with a canal leading from the sea into a harbour basin. On the inland side of the basin the river

[1] B.M., Add. MS. 46348, pp. 609–12; the references are to Bourcher's Mount, Water Mount, Stourton's Mount, Bell or Town Mount, and Berkley's Mount.

[2] B.M., MS. Facs. 372, vol. i, no. 54.

[3] A *new* bulwark implies that there were older works at Ambleteuse to which the new had reference; hence the dating of this plat later than Plats C and D.

[4] *Introduction à la fortification. Troisième partie: suitte des forces de l'Europe* (Paris, 1693), Pl. 5.

entering the harbour is divided into two channels, and each of these is controlled by a sluice gate.[1] This plan was not entirely original with the French, for Rogers and his fellow engineers had developed a similar idea long before, as can be seen by glancing at Plats C and D. That Ambleteuse should become a serviceable port had been a desideratum from the beginning of the English occupation of the town, although it was recognized that there were drawbacks in the natural conditions of the haven. In a friendly chat which took place on 17 April 1547 several English and French officers at Boulogne discussed the merits and demerits of Ambleteuse, and the low tide of the haven was still considered to be one of the major weaknesses of the place.[2] Yet when the French recaptured Ambleteuse in the summer of 1549, Henry II could describe it as 'ung bon grand port, meilleur sans comparaison que celluy de Boullongne'.[3] Exactly what the English had done to improve the haven is difficult to say, since contemporary written documents are lacking in evidence at this point. The three later plats of Ambleteuse all show a similar design for a mole extending into the haven from the sand-hills opposite the fort. Agreement in these drawings suggests that at least this much was done to provide a breakwater for the harbour and to protect the haven against beach drift. But in terms of harbour development the point of real interest on Plats C and D is the barrage extending across the river channel from the fort to a bulwark in the sand-hills on the opposite shore. The dark slots on the up-river side of the barrage probably represent sluice gates. If this is the case, then it will be seen that water from the incoming tide could be trapped behind the barrage by lowering the sluice gates at high tide, or the gates could be used to dam up the normal flow of the river. Once the tide had receded the gates could be raised and the outrush of water used to free ships that had grounded on the shoals of the haven.

This was a commendable scheme, but it remains an open question whether or not it was executed by the English. Henry II's enthusiastic remark about the quality of the Ambleteuse harbour suggests that more was done by the English than building a simple mole. Yet other confirmation is lacking, and the later plat of Ambleteuse fails to show either the barrage or the bulwark on the opposite shore. Even if the project was not

[1] The word *écluse*, written on the drawing twice, guarantees the meaning of the devices shown.

[2] *Cal. S.P. For., 1547–53*, 334.

[3] Rosny, 'Documents inédits', 457.

completed, however, the idea for it may have inspired the similar effort by the French in the seventeenth century.

On 18 February 1547 Lord Stourton, commander at Ambleteuse, reported to the Lord Protector that the bulwarks, curtains, ditches, cellars, storerooms, and gate of the fort were in various stages of completion, and that work was proceeding satisfactorily on most of these projects. The chief difficulty lay in the scarcity of carriages for hauling material.[1] Stourton mentions that Rogers was busily engaged in directing the workers to their specific tasks on the various projects. On 27 March Rogers himself wrote to the Lord Protector that the work at Ambleteuse was getting on reasonably well.[2] But apparently by that time it was not proceeding quickly enough to suit Stourton, who reported around the middle of April that the fort was as yet imperfect, and in particular that the mounts and curtains towards the sand-hills were not high enough to protect the fort from cannonfire from that direction.[3] Stourton did admit, however, that the surveyor was at the moment involved with the completion of the ditches and with preparations for completing the gate, of which there was a great need.

Rogers was indeed a busy man, for besides the main fort under way at Ambleteuse there was a smaller fort being built nearby. The French called this Fort Sellaque,[4] while the English never gave it a special name, but continued to use the awkward circumlocution 'the fort on the hill called the Almaine Camp'. In April 1546 Rogers was instructed to send a plat of 'a bulwark newly to be made upon a hill near this fort'.[5] The next month Hertford was commanded 'to set Rogers in hand with the fortification upon the hill where the Almains [Germans] lie, according to the plat'.[6] This plat apparently has not survived, so that it is difficult to obtain a very precise notion of the character of Fort Sellaque. It is not shown at all on Plats C and D, although the later plat of Ambleteuse provides at least a bare outline in the form of a quadrangular fort with bastions at each corner. This plan agrees with a similar outline for the fort shown on Nicolay's map of 1558, where 'Selacque' can be seen

[1] *Cal. S.P. For., 1547–53*, 305–6.

[2] Ibid. 321.

[3] Ibid. 332. A glance at Plat C will show that these sand-hills were located across the haven from the fort, so that the incomplete curtains were those on the side of the fort parallel to the haven.

[4] In August 1549 Henry II described his approach to the fort: '. . . je feiz marcher mon armée droit devant le fort de Sellaque, prochain de la ville et chasteau d'Ambleteul . . .'. Rosny, 'Documents inédits', 456.

[5] *L. & P.* xxi. i. 686.

[6] Ibid. 884.

adjacent to 'Ambleteul' (Pl. 23). Sellaque does not seem to have been a very momentous undertaking. It was probably entirely of earthwork and was thus similar to Boulogneberg in both form and material. Certainly Henry II, who was much impressed with the English works in the Boulonnais, gave little notice to the fort in his description of the recapture of Ambleteuse in 1549.[1] Even the English themselves seem to have had some doubt about its value, for at the end of May 1546 Hertford noted that the location of the fort was questionable, since it protected satisfactorily neither the haven nor the sand-hills.[2]

The fort which the English called 'Blackness' was located on the Cap Gris-Nez and is clearly shown on Nicolay's map as 'Le blancnes fort'. Work began there somewhat later than at Ambleteuse, for late in May 1546 the decision had not yet been made whether to fortify inland or on the coast.[3] This decision was in time duly reached, but still the work proceeded slowly. In February 1547 it was noted that no part of the wall around the fort had been started, although masons were cutting stones for the cellars and storage houses.[4] By May of 1548 these storage houses were still not quite ready to be used, and the rest of the fort was in even worse condition. The gun mounts and ramparts were reported by Lord Stourton to be in a defenceless state, so that neither ordnance nor men could be placed upon them, while some of the curtains were deficient in height and others were completely disordered.[5] This dreadful condition of the fort may have been overdrawn by Stourton, for he had no great love for John Rogers and was probably colouring his report to discredit the surveyor. Nevertheless, the condition of the fort apparently was unacceptable at this time, and it seems only to have been partially improved in the year or so following. When Henry II appeared before Blackness in August 1549 he remarked that the fortification of the place was 'bien avancée', but this implies that it was not fully completed even at this late date.[6] The form of the work remains unknown from English documentary or pictorial sources, but Nicolay's map shows it as a square or perhaps diamond-shaped fort with bastions at each corner. It seems fairly clear from the English documents that the fort was largely in earthwork, and that the only masonry structures lay in the cellars and storehouses.[7]

[1] See Rosny, 'Documents inédits', 456.
[2] *L. & P.* xxi. i. 960. [3] Ibid.
[4] *Cal. S.P. For., 1547–53*, 306.
[5] Ibid. 342.
[6] Rosny, 'Documents inédits', 457.
[7] *Cal. S.P. For., 1547–53*, 305, 342, 345.

If Rogers did not give as much attention to Blackness fort as Lord Stourton might have desired, he was at least intrigued with the possibility of developing a harbour there. As we have seen, the natural conditions of the havens at Boulogne and Ambleteuse were proving to be something less than desirable. In June of 1546 Hertford waxed enthusiastic about making a harbour at Blackness through the construction of a mole. He sent to Henry VIII a plat of the proposed mole, the idea of which had been worked out by himself, Rogers, and others. Hertford urged upon the king the notion that the resultant harbour would 'do more service for keeping the Narrow Seas than all your havens on this side'.[1] He appears to have been over-optimistic concerning both the plan and the execution of such a project, for in March of 1547 Rogers reported that even a temporary haven was still lacking at Blackness. Unless the project was undertaken, Rogers noted, it would be impossible to use Blackness as a haven, and both building materials and supplies for the garrison would have to be brought overland from Ambleteuse. In the meantime, he suggested building a wall of stone and lime some four hundred feet in length, which would make possible a temporary haven for landing the supplies necessary to build a larger harbour and fort.[2] The Privy Council answered on 2 April that his suggestion for a wall to be used for a landing place would be too expensive and would thereby hinder the progress on the fort itself, which was at the moment the more important of the two projects. Thus he should desist from working on the haven and concentrate upon finishing the fort.[3]

Nevertheless, Rogers continued to study the problem of the haven, and on 7 April he proposed another scheme for Blackness. Although it too never saw any tangible results, we may review briefly the plan which he submitted to the Lord Protector, for it reflects his increasing interest in harbour engineering. His idea was to turn Blackness into an island by making a cut on the landward side of the fort. But we may let Rogers explain the project in his own words:

> It may please your grace to understand that among the things which I have in doing here, as about Boulogne and Newhaven, I am still careful as well for the Blackness and especially for the haven to be made there. And among all my poor imaginations I can think none so feasible as to cut it through the land as your grace hath seen there in the valley that goeth from sea to sea, compassing the

[1] *L. & P.* xxi. i. 1159. [2] *Cal. S.P. For., 1547–53*, 320–1. [3] Dasent, *A.P.C.*, ii. 464.

Blackness in and making him an island, for so shall both the flood and ebb run through the haven and make it always deep. This bearer, a servant of Sir William Godolphin, with the rest that can skill of mining, hath no doubt but they can sink wells to the depth of the low water, which will not be past ten or twelve fathoms one from another, and so to make drifts from the one well to the other beneath—joining them all together save that next the sea, which towards a spring when time is, shall be opened, so that the tide may go clean through, which will with the grace of God do much more in little time than the hands of many men.[1]

Rogers suggested that this work could be executed in six months by one hundred miners. But there was little hope that the English Government could undertake this project, for it was already more deeply committed to the other expensive works in the Boulonnais than it could well afford. If funds could not be found for building the fort at Blackness, what chance was there for this scheme to turn Blackness into an island? This plan projected by Rogers was therefore of little immediate consequence, but it is interesting to note that the idea of a 'cut' would later be used by Rogers when he was called to Sandwich, Kent, to deal with a special problem of harbour engineering there.

[1] P.R.O., S.P. 68/14, no. 103 (*Cal. S.P. For., 1547–53*, 327).

V

A MAN WITH MANY DUTIES

As we have moved with Rogers to his assignment in Boulogne, the biographical element in our study has become more prominent, for the details of his career there are much more fully available than for the preceding years. It is still impossible to draw a full-scale portrait of the man. For instance, we have not the slightest idea of his physical appearance—whether he was tall and thin or short and fat, whether bushy-haired or bald-headed. But something of his character does begin to emerge from the correspondence in the state papers, and it turns out to be a rather interesting character at that. Not only does Rogers's personality become more clearly defined; his relationships with Henry VIII, with his superior officers, and with his fellow engineers unfold in the voluminous correspondence dealing with military fortifications at Boulogne. As a result, incidentally, we learn some interesting things not only about Rogers, but also about the English king and his ministers. Furthermore, through an examination of the details of the work at Boulogne, the role and significance of the sixteenth-century surveyor of works become explicit in the activities of a particular surveyor, thereby enlivening the somewhat abstract definition of that office which we have thus far produced.

The first question which arises in regard to that office is to what extent Rogers, as surveyor of works, was responsible for designing the overall scheme at Boulogne as well as the individual projects. In order to arrive at a satisfactory solution to this problem, a number of considerations must be kept in mind. As previously noted, the overall plan at Boulogne seems to have developed on an *ad hoc* basis—with new works projected as the need arose, rather than a grand scheme having been devised by one or several minds and then executed in its various parts. Secondly, the surveyor of the works was considered at Boulogne to be the chief technical officer connected with the building projects and was *ipso facto* the artificer or deviser of those works. But he was to some extent limited in the exercise of this function by the chief military officers at Boulogne, as well as by the king and his ministers, many of whom had definite opinions on what fortifications were needed, and where and how they should be built. Henry VIII in particular had made up his mind on a number of points during his few days' sojourn in Boulogne during the autumn of 1544, and he continued to develop other ideas as the works proceeded. Sometimes these ideas were sufficiently precise to be presented in his own plats of proposed new projects, such as that at Ambleteuse in the spring of 1546. More often Henry reserved for himself the right to correct or amend the plans presented by his officials. There can be little doubt that in this way the king played an important part in the final determination of the form and character of individual works. Likewise the military officers in the Boulonnais sometimes affected the course of the new work, particularly if they considered themselves to be knowledgeable in matters of military architecture and engineering. The most influential officer in this respect was the military governor of Boulogne, to whom Rogers as surveyor was subordinate. William Lord Grey held this office from April 1546 to August 1547, and during this period he figured prominently in both planning and executing the works; indeed, his tenure was the most productive period in the history of the English works at Boulogne. In addition, each fort had its own captain, who might actively engage himself in the construction of the fort under his command.[1] William Lord Stourton, captain of Ambleteuse, was particularly prominent in this regard. Finally, the part played by special technical advisers, such as Sir Richard Lee and John Burgate, should not be overlooked.

Thus the design of a new work was often the result of a co-operative endeavour, in which an initial plan might be devised on the spot by the surveyor in conjunction with one or more advisers, both military and technical. This plan would be submitted in the form of a plat to the king and his ministers, who would make changes or approve the plan as presented. Work on the project could then get under way.

When Rogers was initially appointed surveyor of works in September 1544, he was in a sense hedged in with military and technical advisers who could both assist and check him in his work. Then shortly afterwards the work at the Old Man was removed from his jurisdiction and placed under the direction of Sir Thomas Palmer. But Henry's confidence in Rogers's

[1] Cf. the scheme for the government of Boulogne which lists the various types of officers and ministers at each of the forts in the Boulonnais for the year 1546: *L. & P.* xxi. i. 505.

ability and judgement increased rapidly during the following year as a result of the exchange of ideas through letters, plats, and personal interviews which the king and his surveyor enjoyed. By March of 1546 Rogers's stock had risen so high that he stood alone in charge of all the works at Boulogne. This point was made explicit at the time in an interesting letter from William Paget to the Earl of Surrey. Paget explained that the king had originally given to Surrey, Palmer, and Rogers the responsibility of ordering the fortifications. But Surrey and Palmer had shown such uncertainty in their opinions that the king had conferred with Rogers, had devised certain plans with him, and was now placing the execution of those plans entirely in the surveyor's hands. Recognizing the implications of this move, Paget warned Surrey and Palmer:

> . . . concerning the opinion both of your lordship and the said Sir Thomas Palmer, that neither of you will for this cause or any other thing tending to the advancement of his majesty's service or profit bear the worse will to the said Rogers, but rather favour him and support him accordingly. Your lordship knoweth the man is plain and blunt, which must be borne withall, as long as he is well-meaning and mindeth the service of the king's majesty . . .[1]

It was necessary for Paget to add this last bit of advice, for the plain and blunt ways of Mr. Rogers had already got him into more than one fracas with the military officers at Boulogne, and in these disputes the officers found to their dismay that Henry was as likely to side with Rogers as with them. Rogers became aware of this, and the record of his surveyorship at Boulogne shows him frequently at odds with his administrative superiors, seemingly in full confidence that his opinion would prevail. This arrogant attitude, needless to say, produced considerable ill-will against the surveyor on the part of the Boulonnais officials. As early as January 1545 Lord Lisle complained to the Council that there was no provision of building materials, that Rogers had returned to England without letting his intentions be known, and that even his wardens and clerks had departed without informing the proper officials. As Lisle ruefully remarked, none of the necessary works could be ordered in the absence of the surveyor and his subordinates.[2] From this incident, and from another that occurred in June of the same year, it is clear that Rogers guarded carefully his prerogatives as technical supervisor of construction. It appears

[1] P.R.O., S.P. 1/215, f. 35v (*L. & P.* xxi. i. 356). [2] *L. & P.* xx. i. 39.

that Richard Cavendish, the comptroller at Boulogne, was attempting to exercise some authority over the construction workers. He was vehemently opposed in this by Rogers, who claimed that when he was last at Court he was told by the king himself that Cavendish should not meddle with the labourers.[1] Presumably Rogers won the point, but Cavendish got his revenge a year later when he was placed in charge of the works at Blackness and Rogers was specifically told not to meddle there.[2]

Rogers was not content merely to test the authority of comptrollers like Cavendish; he would take on the very governor of Boulogne itself, William Lord Grey. In a letter to the king, Grey explained the incident.

> Albeit I have been very evil used oftentimes heretofore by Rogers, your highness' surveyor here, as well in his overthwart taunts and querulous language, with comparison of his service to mine, as otherwise in such sort as he hath neither taken me as his equal nor his superior—according to the honor of the place that it hath pleased your majesty to appoint me—but even as I were his inferior! Yet trusting to have won him by gentle means, I have borne his weakness until now that in like manner he chargeth me with untruth to your highness, which I cannot in anywise bear nor suffer.[3]

It seems that when the farrier smith ran out of coals for his furnace, Grey requested some from the stockpile in Rogers's charge, offering either to pay for the coal then or to replace it when other coal arrived. According to Grey, Rogers retorted in the presence of several people:

> Show my lord that I am not made purveyor of coals, neither for him nor the rest. And he shall use things more truly before he shall have anything at his commandment that is in my charge. And tell him further that I myself begin now to ruffle, to the intent that I will have other things known.

This was rather strong language for a mere commoner—even if he was the king's surveyor of works—to be directing against his lordship. Consequently, the Privy Council sent instructions that Rogers was to be called in and reminded of his duty to his chief officer. He was to be told that

> like as the Kinges Majestie toke hym to be a good servaunt and did favour him for his well doinges, semblablye it semed to his Highnes to be thoffice of a good servaunt not onely to doo well, but also to saye well . . . and that it is not ment that he whome his Highnes hath there constitute for his chief officer shuld be with wordes unfitting or uncommely behavour seme brought in to contempt . . . wherfore his Highnes

[1] Ibid. 884.
[2] Ibid. xxi. i. 1234.
[3] S.P. 1/221, f. 53 (*L. & P.* xxi. i. 1172).

wold he shuld amend such his fault, and soo texecute his duetie in the service of his Majestie committed unto him, as also he might have regard to use his wordes more discretelye unto those whome his Highnes in respect of their placeng wold to be regarded accordinglye.[1]

In spite of this warning, Rogers continued to speak his mind in no uncertain terms to the officials at Boulogne, and it displeased some of them greatly to have to tolerate the man. It is revealing to watch Nicholas Arnold, captain of Boulogneberg, try to keep his language in check when referring to Rogers. In January of 1547 Arnold wrote a letter explaining the unhappy physical condition of his fort. There was no well, no place to store food or munitions, no gate or drawbridge, and the ramparts and bulwarks had been ruined by the frost and rain. He complained that it was dangerous to leave his fort in that condition, but then he added, 'lest I should seem to seek my own safety rather than the King's service, "as Mr. Rogers, our gentle surveyor here, hath sundry times told me that I do when I have sought redress at his hand for the amendment of some of these things," I remit the matter to your wisdom'.[2]

Rogers's most consistent opponent was William Lord Stourton, captain of Ambleteuse. On 2 January 1547 Stourton complained that at his fort there was no place to store grain and no bakehouse, and he blamed the surveyor for refusing to correct the situation.[3] The Privy Council suggested in response that 'a certain long house' be used for storage space. Stourton replied that Rogers was using one part of that house as a brewery and the other part as an arsenal—in spite of the fact that he had warned Rogers several times against keeping gunpowder so near the brewhouse. But, adds Stourton, since the surveyor has built for himself a handsome house in which he has never slept, as well as similar houses for the master mason and master carpenter, he will confiscate these for storage space until further provision can be made. One can almost hear the grinding of lordly teeth as Stourton concludes: 'And for myself to speak to the Surveyor for anything touching the strength of the fort or anything else, he giveth me none other than a mock or a taunt, which I would be loath to suffer at his hands, or at a far better man's hands than he, but only for fear of the King's Majesty's displeasure.'[4]

Stourton knew very well that he would incur the king's displeasure if he

[1] Dasent, *A.P.C.* i. 474–5.
[2] *L. & P.* xxi. ii. 731.
[3] Ibid. 655.
[4] Ibid. 667.

misused Rogers, but in less than a month after this letter was written Henry VIII was dead, and Rogers's position became more precarious. The change is reflected almost immediately. On 7 February 1547 Stourton wrote to the Privy Council that he had repeatedly sent for the surveyor, who had five times failed in his promise to come, and who had on that very day informed Stourton that he did not know when he would be able to come. The worthy captain had then proceeded to remove the workers from the task assigned by Rogers and to put them on another project. If, says Stourton, the surveyor feels himself aggrieved at this move and complains to the Council, they will please consider Stourton's superior reasons for controverting the surveyor's plans.[1] On various occasions during the next year Stourton criticized Rogers's handling of the works at Ambleteuse and Blackness, and in time those criticisms began to tell. By May of 1548 Rogers was clearly on the defensive, as indicated by the last extant letter from him, in which he tried to explain to the Privy Council the work that he had been doing. He remarks somewhat pitifully, if not petulantly:

> But it is little wonder though I being but one a poor man should not be able to please so many idle as here are doing nothing but imagining to trouble me. . . . Other accounts have I none to make, but my heart is not a little heavy that my doings doth no better please your good lordships, I having no trade nor occupying in this world but the king's majesty's business, wherein I travail night and day—and yet so little accepted, as though I did my own business and left the king's. Alas, how little joy brings this into my poor heart and how great heaviness have I cause to be in, to consider all this.[2]

From the time of Henry's death in 1547 Rogers's influence in affairs at Boulogne was on the wane. Although he remained as surveyor of works until the end of the English occupation in 1550, he never enjoyed the confidence of Edward VI, the Lord Protector, or Edward's other ministers as he had enjoyed that of Henry VIII during the later years of that monarch's reign. Perhaps this was the revenge of officials in Boulogne and London for the insolence which Rogers had displayed when he was under Henry's protection.

If Rogers's conflicts with the officers at Boulogne eventually led to a restriction of his powers, nevertheless he continued to be one of the most active officials in the Boulonnais. Our description of his scrapes has itself shown that his opinion on this or that problem was often sought and cited,

[1] *Cal. S.P. For., 1547–53*, 298. [2] S.P. 68/15, no. 141 (*Cal. S.P. For., 1547–53*, 340).

and that his direct supervision was generally considered to be necessary for the satisfactory progress of the work. When one reflects on the fact that in the summer of 1546 there were at least nine separate projects under way in the Boulonnais—all of them more or less under the immediate supervision of Rogers—one may well imagine that he was indeed a busy man. Probably he had sufficient reason for failing to heed the beck and call of Stourton, who was after all concerned with only a part of the whole operation which Rogers had to keep constantly in mind.

Then there must be added to Rogers's tasks of planning and supervision the many other functions which he fulfilled as surveyor of works. For instance, he was often involved with the problem of obtaining labourers for the works. In May and November of 1545 and in April of 1546 he was issued warrants and money for impressing and transporting workers from England.[1] This involved personal trips away from the works at Boulogne, which must have helped to create some of those difficulties with Boulogne officials that have been noticed. But the acquisition and retention of a sufficient number of labourers was one of the major continuing problems at Boulogne, and it probably accounts for the fact that Rogers himself sometimes took a hand in the matter. Although masons, bricklayers, and carpenters were needed for the works, the main brunt of the labouring force consisted of pioneers—that is, men employed in constructing earthworks. Most of these labourers could be relatively unskilled, and there is evidence that they were lowly paid or even *un*paid at times, because of lack of funds in Boulogne.[2] When this was combined with the poor living conditions for the workers, it is not surprising that Rogers and the other officials found it difficult to retain these men.[3] In January of 1545 there were 1,200 pioneers working at the Old Man; by June so many of them had died or gone away sick that only 300 remained, while in July the number dwindled to barely 100.[4] Urgent pleas were sent for pioneers for this important work, and apparently the need was finally realized at Westminster, for by the end of August more than 2,000 pioneers had been

[1] *L. & P.* xx. i. 557 (f. 58) and 821; xxi. i. 643.

[2] Ibid. xx. ii. 558; *Cal. S.P. For., 1547–53*, 331.

[3] Cf. *L. & P.* xix. ii. 500, 683; xxi. ii. 82; and *Cal. S.P. For., 1547–53*, 331.

[4] It is interesting to note the attitude of the Privy Council towards those who happened to be present at Boulogne during this time of great sickness. On 25 June a letter was sent to Lord Poynings 'to take ordre in respect of the Plage att Bullone, that none suche as shall repayre hither from thens shall adventure to coome to the Courte untill suche tyme as they shall have been abrode and ayred'. Dasent *A.P.C.* i. 202.

assembled at the Old Man. Yet this number had dropped off sharply again by March 1546.[1]

A partial solution to this dearth of pioneers was to conscript the soldiers garrisoned in Boulogne for work on the fortifications. This was attempted as early as November 1544, when the need was great to make at least temporary repairs in the earthworks of Basse Boulogne before the onset of winter.[2] Grumbling on the part of the soldiers was the inevitable consequence of this decision, at a time before the pick and shovel had come to be accepted as part of the equipment and fate of the infantryman. But the soldiers' ruffled feelings were assuaged with extra pay for extra duties. The Earl of Hertford seems to have been particularly skilful at getting this kind of service from his men. Yet their enthusiasm for the work which he reported in June 1546 was perhaps due more to his promise that the crews which finished their projects first would be the first to return to England than to the promise of a groat a day for each soldier's efforts.[3] Furthermore, in view of the zeal for their work, it is doubtful that the soldiers realized what Hertford himself knew, namely, that the money for this extra pay was not yet available. In February of 1547 Edward Dudley, appointed as captain of the Young Man, likewise promised extra pay to his soldiers for their labour, and so anxious was he for the completion of the fort which he would command that he used some of his own money to meet the expenses of fortification.[4] As surveyor of works Rogers must have been grateful for this kind of zeal and co-operation on the part of military officials. This is evident in a letter concerning the Young Man, where he mentions Dudley's enthusiasm and then remarks that he is counting on the captain's men to dig the cellars and foundations of the fort.[4]

Besides participating in the acquisition of labourers for the works, Rogers also played a role in the purchase of materials and in the accounting of expenses for both labour and materials. All of these functions are neatly outlined in a paragraph of one of the few building accounts surviving from the works at Boulogne:

Also paid by the said accountant for the wages of diverse artificers and labourers working upon the king's fortifications within the High Town, the Base Town, and the Young Man, under John Rogers, surveyor of the works there; for diverse

[1] *L. & P.* xx. i. 121, 925, 1160; xx. ii. 200; and xxi. i. 438.

[2] Ibid. xix. ii. 683.

[3] Ibid. xxi. i. 1133.

[4] *Cal. S.P. For., 1547–53*, 306.

empcions [purchases, supplies] bought by him in England and Flanders for the same works, and for freights of diverse provisions coming out of England from the 20th of December, anno 36 [1544] to the 24th of September, anno 37 [1545], by the space of 12 months 27 days, accounting 28 days to the month, as appeareth in 13 particular books subscribed with the hand of the said John Rogers . . . in all £10,402. 7*s*. 4*d*.[1]

Another account indicates that during the first year of the works at Boulogne, from September 1544 to October 1545, the wages for the labourers alone came to £20,875.[2] Unfortunately, expenses are unavailable for the big building season of 1546, when Clement's Bulwark, the town walls of Basse Boulogne, and the forts at Boulogneberg, Ambleteuse, and Blackness were begun, but the costs must have been considerably greater for that year. It is little wonder that from 1547 onward the Government of the Protectorate, which had economic troubles enough, found it increasingly difficult to pay for the continuation of the many projects at Boulogne which Henry VIII had begun. Nor was Rogers in an enviable position as he tried to bring to completion these several projects with inadequate supplies of men, materials, and money. Add to these difficulties the impatience of the local commanders, who wanted their particular fortresses finished in a hurry, plus Rogers's apparently natural acerbity, and one can well see why tempers flared and conflicts occurred between the surveyor of works and some of the military personnel at Boulogne.

Most of these duties which Rogers was performing at Boulogne had traditionally been associated with the chief technical official of the works, whether he had been the master mason, the master carpenter, the master of the works, or the surveyor of the works. But at Boulogne Rogers became involved in one kind of activity which had not previously been a part of the English building technician's functions. In the spring and summer of 1546 he played a quite important role in the diplomatic negotiations which preceded and followed the Treaty of Camp—signed on 7 June—which ended the war between England and France. Rogers's part in these negotiations was that of a special kind of diplomatic attaché, whose functions were to survey the territories in dispute and to provide topographical drawings which could serve as a basis for negotiations on these territories. It was, therefore, his fame and skill as a platmaker which drew him away

[1] P.R.O., E. 101/207/1.

[2] *L. & P.* xx. ii. 558. Characteristically enough, of this sum only £17,488 had been paid at the time of accounting, while £3,387 in wages remained unpaid.

from supervision of the work in Boulogne and got him involved in international diplomacy. But it was also Henry's confidence in Rogers's general ability that occasionally earned for him an important place in the actual negotiations. Finally, because of his familiarity with the land surveyed, he was sometimes used as a liaison between the English diplomats in France and the king in England. When he could not travel, his plats carried information back and forth across the Channel. It should be noted, however, that Rogers was not the only building technician involved, for the surveyor of Calais also participated at times in the negotiations, as did Sir Richard Lee. Indeed, 'Rogers and Lee' formed a kind of team in making surveys and producing plats therefrom, and in conferring with the English diplomats and Henry VIII on matters that called for their specialized knowledge and skills.

Rogers must have got his introduction into this kind of activity as a platmaking spy of enemy fortifications. He performed this function at Guines in 1540 and in Scotland in 1544, and these occasions no doubt revealed his usefulness to the king in this role. In July 1545 he was again placed in this role when he was directed to make a plat of the French camps and fortifications on the south side of the Liane River.[1] Then in May 1546 Rogers was requested to produce a plat for consideration in the diplomatic negotiations for ending the war between the English and the French. For these negotiations, which were just then seriously getting under way, Henry sent to his diplomatic commission at Calais—consisting of Lord Lisle, William Paget, and Nicholas Wotton—the following instructions. The English must be left in possession of Boulogne and part of the Boulonnais and the county of Guines until the French had paid certain arrears and debts, in addition to two million gold crowns indemnity. The territories in question would be signified by a plat which they would receive as soon as Rogers, who had been sent for by the king, 'may come and go unto you'. If the French refused to accept the terms of peace as stated in the letter and shown on the plat, but were yet willing to make a truce, then a truce might be made—provided that the king retained the above-mentioned territories as shown on *another* plat to be sent by Rogers.[2] On 11 May Paget acknowledged this letter and the promise of the two plats, and urged that the plats be sent immediately, since knowledge of the territorial limits as defined by the drawings was necessary for his next

[1] *L. & P.* xx. i. 1192. [2] Ibid. xxi. i. 775. Letter dated 9 May.

meeting with the French.[1] Hertford wrote to the king from Newhaven on 13 May that he had sent Rogers with a plat of the country. On the same day Paget acknowledged the arrival in Guines of Mr. Lee and Mr. Rogers, 'who hath viewed the hills and showed me somewhat whereby we may work touching that point', the point presumably being the establishment of peace.[2]

When the peace commissioners on both sides met, the French agreed to most of the stipulations set forth by Henry VIII, with the proviso that Henry would retain the Boulonnais on the northern side of the Liane only up to the town and forest of Desvres.[3] On 17 May Henry replied that this agreement was satisfactory, but that since he did not know where the 'river Devre' begins or flows, this point would need verifying before a formal agreement could be reached. He suggested that Paget with one of the French commissioners should survey the river, secretly accompanied by Lee and Rogers, who were being sent to assist Paget in this matter.[4] After the survey had been completed, Rogers should return to his duties in Boulogne, while Lee should bring a plat, 'the making of which need not be intimated to the other side'.[5] Paget replied on 19 May that the French commission was too impatient to wait for the English to survey the river, make a plat, send it to England, and receive an answer in return.[6] Nevertheless, two days later the French agreed to the survey, if all other points were agreed upon. At the same time they presented a plat of the country, the accuracy of which, they said, could be tested at 'the treading of the ground'. Meanwhile, Lee and Rogers appeared upon the scene, but before they could execute the proposed survey Lee was sent to the king with a plat of the country, which seems to have been the one that the French had produced.[7]

By 27 May the principal points for negotiating the peace had been settled, but there was still considerable controversy over the question of the boundary of the Boulogne Pale as defined by the upper reaches of the Liane River. The main desire of the English seems to have been that of obtaining a boundary on the east which would connect Calais, Guines, and Boulogne, and which would incorporate them into one pale, so to speak.

[1] *L. & P.* xxi i. 792.

[2] Ibid. 817, 818.

[3] Ibid. 841. These are shown on Nicolay's map (Pl. 23) as the town and forest of Desurene.

[4] Apparently Rogers and Lee had departed from Guines after their arrival with the plat on 13 May, but now they were on their way back to the town.

[5] *L. & P.* xxi. i. 849.

[6] Ibid. 862.

[7] Ibid. 871.

On the other hand, the French feared that an extension of the English boundary too far eastward would block their own passage between Montreuil and Ardres, their two strongest fortress towns just outside the English-held territories. In short, the French were finding themselves squeezed between the English holdings on the coast and the Hapsburg-controlled County of Artois, while the English felt they needed to secure their possessions by connecting them all within a single boundary.[1] It was for these strategic reasons that the long and frustrating negotiations over the boundaries continued, rather than because of mere English satisfaction in holding a few more square miles of French land, or wounded French pride determined to keep the English from getting those few square miles.

One of the problems which the English faced was their lack of detailed topographical information on the area, and to correct this they sent forth a surveying team to find the head of the river. This expedition failed, but a second one surveyed the area and made a plat of the country.[2] Still the haggling over the boundaries continued. As late as 2 June Henry informed his commissioners that the uncertainties regarding the boundaries, including those between the head of the Liane River and Guines, should be settled before a final agreement could be reached; he suggested that Rogers be sent to survey those boundaries once again and to make a plat of them.[3]

On 7 June 1546 the negotiations finally produced the Treaty of Camp, bringing an end to the war and a settlement of the major issues. Regarding the vexed question of boundaries it was agreed that the English would hold the port of Boulogne, which was defined to include the river and its south shore to the point where the highest winter tides reached. From the haven the boundary was to extend along the south shore of the river up to the Pont de Brique, from which point the river would be common to both sides up to its head and font (*ad Caput & Fontem*).[4] The precise location of this 'head and font' was not spelled out, nor were the boundaries from there to Guines defined; these sticky issues were intentionally left to be resolved *after* the treaty had been established.[5] Thus, early in August English and French commissioners were still trying to determine the proper head of the river. By this time the problem had been considerably

[1] Cf. ibid. 941.
[2] Ibid.
[3] Ibid. 974.
[4] Thomas Rymer, *Foedera*, 2nd ed. (London, 1728), xv. 97; cf. *L. & P.* xxi. i. 1014.
[5] *L. & P.* xxi. i. 1427 (2).

refined. It centred upon the question of which one of the two branches that met near Selles should be considered as the source: the one which rose near Quesques or the one near Vieil-Moutier? When it was decided that the longer branch should be reckoned as the head, the problem resolved itself into a matter of quantitative measurement. For this reason the surveyors of both Calais and Boulogne took part in the negotiations. But their platmaking abilities were also utilized, for the question of the boundary between the head of the river and Guines was not merely a mathematical one, but a political one as well, since it involved lands that touched upon those belonging to the emperor. Further diplomatic negotiations were necessary; regarding these, Henry instructed his commissioners to keep him posted on the proceedings, while the surveyors should send plats from time to time showing the differences between the English and the French claims.[1] In spite of the time and energy already devoted to the question of the boundaries, and the clarifications which had resulted, the problem was not quickly settled. Negotiations continued, and as late as November 1546 Rogers was appointed to another commission, whose task it was to solve the old riddle about the head of the Liane River.[2]

Throughout these negotiations between the French and English, Rogers had been requested time and again to make plats for the perusal of the commissioners and king, and there is documentary evidence that he did in fact produce several drawings. In several ways Survey I of Boulogne (Pl. 19) fits very nicely the reference to the plat of the French camps and fortifications which Rogers produced in July 1545.[3] Although the drawing does indicate the works on the English side of the haven, it is obviously more concerned with the French fort and camps on the south side of the river, for it is to these that the fairly lengthy explanatory inscriptions are applied. The details of the plat also correspond with the above-mentioned date, for both the French and the English works are shown at the stage which they had reached by midsummer of 1545. Yet the attribution of the drawing to Rogers faces serious difficulties which arise from considerations of the drafting technique of the plat. The overall impression given by this drawing is quite different from that produced by other contemporary drawings attributable to Rogers (Pls. 15–17, 20–21). This first impression is substantiated by a closer look at the details, where fundamental differences in drawing technique can be observed. The draughtsman of Survey

[1] *L. & P.* xxi. i. 1427. [2] Ibid. xxi. ii. 422. [3] Ibid. xx. i. 1192.

I has produced a free-hand drawing, in which most of the straight lines, circles, and curves were sketched by means of very short pen strokes. On the other hand, the straight lines, as well as most of the curved ones, on the plats attributable to Rogers were made with a continuous stroke of the pen from the beginning to the end of the line. Furthermore, a straight-edge was used on these drawings even for short lines, instead of the free-hand technique largely employed on Survey I. Because of these basic differences, it seems wiser to conclude that some draughtsman other than Rogers produced Survey I of Boulogne, even though the contents of the drawing correspond so closely with the reference to a plat produced by Rogers in July 1545.

The details of the various works shown on that plat which we refer to as the 'Country of Guynes and Bolenois' (Pl. 14) suggest that it dates from the spring of 1546. High Boulogne, Basse Boulogne, the Young Man, and the Old Man are at their proper stages of development for this date, whereas the works at Boulogneberg, Ambleteuse, and Blackness are not indicated, since they were begun late that spring. It appears very likely that this was one of the several plats made during the negotiations in the spring of 1546, preceding the Treaty of Camp. In fact, this particular plat may be the subject of that series of letters written in May, in which both Henry VIII and Hertford noted that Rogers was on his way to Guines with a plat of the country thereabouts.[1] On 27 May William Paget reported that if the French determination of the head of the Liane River was accepted, the king would lose the village of Lottingham, 'which in that chart that Rogers brought we think be not truly set'.[2] Now, Lottingham (i.e Lottinghen) is shown on the plan under consideration as 'Lowting'. It is placed on the further side of the river, away from the English-held territories. But the river is mistakenly shown here as a single stream from its source, without the two branches which are in fact in its upper reaches. Furthermore, the village of Lottinghen actually lies between these two branches, one of which begins near Quesques, the other near Vieil-Moutier.[3] Since the English were claiming Vieil-Moutier as the head of the river, Lottinghen would lie on *their* side of the river, rather than in French territory, as Paget was explaining. Thus his statement that

[1] Ibid. xxi. i. 775, 817, 818.

[2] Ibid. 941.

[3] The two branches are clearly indicated on Nicolay's map of 1558 (Pl. 23). 'Quesque' is shown, though Vieil-Moutier is not, and Lottinghen is seen as 'Lobingen'.

'Lottingham' is misplaced on the plat which Rogers brought closely corresponds with the contents of the present drawing. The internal evidence of the drawing also points to Rogers. For instance, the curious roofless houses are drawn in the same manner as those on Plats A and B of Ambleteuse (Pls. 15 and 16), while the drafting techniques of the three plats show strong similarities. The inscriptions on the plat were not made by Rogers, but the endorsement on the verso, 'A plat of the contre of gy[nes and] bolinnoys', appears to be in Rogers's hand.[1] When all this documentary, circumstantial, stylistic, and handwriting evidence is brought together, it seems fairly safe to conclude that this plat was drafted by Rogers.[2]

On 13 June Hertford wrote to Henry that, in response to the king's request, Rogers had made a plat of the country around Boulogne and was at that moment repairing to England with the drawing.[3] It is possible that the drawing to which he referred is Survey II of Boulogne (Pl. 20), which shows the English and French works as they appeared during the summer of 1546.[4] But there is a curious problem connected with the red dotted line on this plat.[5] It might at first be thought to show the boundary between the English and French territory at Boulogne as established in the Treaty of Camp, where the southern limits of the haven were set at the high-water mark of the winter tide. Yet the dotted line could hardly conform to that mark, since it is placed south of the village of Outreau, which would have meant the inundation of the village at high tide. Besides the improbability of that situation, two of the contemporary plats of the area (Pls. 14 and 19) show the haven at high tide, with Outreau perched safely above the water. Another difficulty with the dotted line is that at a point a few hundred feet inland from the shore it turns sharply and runs southward, separating Fort d'Outreau from the village of Le Portel, which the English called 'Pawlet'. If this was intended as a boundary line, it would scarcely have been acceptable to the French, since it would have cost them their corner post overlooking the haven of Boulogne, as well as the village of Le Portel,

[1] See Appendix A.

[2] However, it may have been drawn by Richard Lee, who was working closely with Rogers at this time. Until criteria for distinguishing between Rogers's and Lee's plats have been produced by a thorough study of Lee's handwriting and drawing styles, categorical assertions about the authorship of this particular plat should not be made.

[3] *L. & P.* xxi. i. 1055.

[4] That the works at this time had reached the stage shown on the drawing can be ascertained by comparing carefully the drawing with the history of the works as we have derived it from documentary evidence.

[5] This red line is a prominent feature on the original plat, though it does not appear so distinctly in a black and white reproduction.

which they used for bringing in supplies from the sea. But for these very reasons the English would have liked to establish the boundary line as shown on this plat. To have done so would have been to make good the error committed in October 1544 when they failed to secure the shore of the haven opposite the Old Man. Thus the rationale behind the dotted line is easy to perceive, although the particular circumstances of its conception must remain conjectural. Does the line represent a claim put forward by the English during negotiations, which they later compromised in the settlement of the Treaty of Camp? Or could it be a goal (or perhaps merely wishful thinking) which they retained even after signing the treaty? The particulars of that agreement were cynically treated by both sides, and as we shall see in analysing Survey III of Boulogne, the English were still toying with the idea of securing a foothold across the Liane from Boulogne. The presence of the line, therefore, cannot be used to date the drawing before or after the Treaty of Camp, since it could have been conceived on either side of that event. Indeed, it might have been added subsequently by a hand different from the one which initially drew the plat—perhaps by some minister of state, or Henry VIII himself, who was musing over ways of protecting the English enclave at Boulogne.

If the puzzles connected with the red dotted line remain unsolved, there is at least greater certainty regarding the authorship of the plat as a whole, for besides the circumstantial evidence of Hertford's letter of 13 June, the handwriting and stylistic evidence of the plat point to John Rogers as its draughtsman. The inscriptions were fairly certainly written by the man whom we have called Clerk C and whose handwriting is known from a letter which he scribed for Rogers in March 1547.[1] This is the same hand that wrote the inscriptions on Plats A and B of Ambleteuse, as well as Survey III of Boulogne (Pls. 15–16, 21). Furthermore, the drafting techniques of Surveys II and III of Boulogne are almost identical, as are the techniques of Plats A, B, and C of Ambleteuse. Between the two sets of drawings there are sufficient similarities of style—in the drawing of cliffs, beaches, and rivers, and particularly of the odd roofless houses—to suggest strongly that they were all drafted by the same hand. Thus, if certainty regarding the authorship of any one of the drawings can be achieved, it follows that the draughtsman of all five plats is known. This is why Rogers's monogram on Plat B of Ambleteuse constitutes such an unusual and

[1] See Appendix A; the letter is partially reproduced in Pl. 28.

important clue, for it guarantees the identity of the draughtsman not only for this drawing, but for the other four drawings as well.

Survey III of Boulogne may be ascribed to Rogers because the drafting style and handwriting are so similar to those found on the other plats of Ambleteuse and Boulogne that were drawn by Rogers. However, there is relatively little documentary evidence which connects Rogers with Survey III or which relates to the problems of the date and purpose of the drawing. As a result, these questions must be answered largely on the basis of the internal evidence of the drawing. It would appear from the elaborate network of fortifications shown on the French side of the river that the drawing was executed rather later than June 1546—the approximate date of Survey II of Boulogne. Strangely enough, however, this drawing does not show the Young Man (which is shown on Survey II), nor does it indicate the Dunette, which was begun on 1 April 1547. On the other side of the river, the French project next to the sea—Fort Châtillon—is indicated, although that fort was not constructed until after May of 1548.

The explanation of these puzzling features is fairly simple. Survey III is a plat executed by Rogers in August of 1546, in which he projected the completion of yet unfinished works and suggested others which had not even been begun. The date of the drawing can be determined by the dating of Fort Châtillon, the fort next to the sea that was begun the first time by the French in July 1546, but was destroyed by the English on 5 September. Not until May 1548 was the fort successfully begun again.[1] Thus this drawing must show Fort Châtillon as Rogers could project it either in the late summer of 1546 or in May 1548. But by the latter date the Dunette was well under way, and Rogers surely would have shown it rather than the older French pile-jetty which is seen on this drawing. Consequently, the above-proposed date of August 1546 is the preferred one. This conclusion is neatly supported by the circumstances connected with the Young Man, for it was during this month of August that the fate of the work hung in the balance as to whether it should be abandoned or continued.[2] When Rogers was drafting Survey III he apparently assumed that it would be abandoned, and he has shown this by omitting the Young Man from the drawing and by replacing it with a bastioned trace from the Old Man to the Citadel. But in fact the decision was made by 9 September to continue

[1] Cf. Pierre Héliot, 'Les fortifications de Boulogne sous l'occupation anglaise (1544–1550)', *Revue du Nord*, xl (1958), 35.

[2] See above, p. 68.

the fortification of the Young Man, and by 1 December work was again under way there.[1] Rogers therefore must have drawn Survey III before that decision was made in early September. A final confirmation that the drawing was not executed after August 1546 lies in the fact that no French works are shown at the village of Pawlet or at Pawlet Hill on the sea coast, although the French began work there at the end of August. In fact, on 28 August Rogers and Sir John Brydges were sent to spy on the new work, and they reported that eighty to a hundred pioneers were already at work.[2] If Rogers had known of this work when drafting the plat in question, surely he would have included it in his drawing.

While the date of the drawing has been ascertained only by careful analysis of the internal evidence and a few supporting documents, the subject of the plat is plainly stated in an endorsement written by the same hand that wrote the inscriptions on the face of the plat. It reads 'of Boullen w^t a devyes of a campe for the wynnyng of the frenshe fortyfycacion foranenst bullen'. This device can be seen at the approximate centre of the drawing, where a bulwark is mounted with two cannons which are firing into the French Fort d'Outreau. Beneath this bulwark is an inscription, 'The place to make a mownte', while above it another inscription reads 'The camp next the medowe'. The rather elaborate scheme of English counter-fortifications depicted on this plat represents a logical extension of the idea expressed by the dotted line on Survey II: where the boundary line would have limited French interference in Boulogne haven, the device would have removed the French entirely from the scene. Whether Survey III constitutes a serious proposal by Rogers to build fortifications on the French side of the river for such a purpose, or whether it was merely an idle 'war game', I do not know.

Two other drawings relating to Boulogne may be briefly mentioned. It will be recalled that Sir Richard Lee had returned from Guines on 21 May without completing his and Rogers's mission of surveying the Liane and making a plat of the country. He did, however, return with a drawing supplied by the French commissioners. At the time William Paget described this as the fairest and best plat of the Boulonnais he had ever seen.[3] It seems reasonable to identify this plat with Survey IV of Boulogne (Pl. 22), which from the state of the fortifications as depicted indicates that it

[1] *L. & P.* xxi. ii. 54, 482. [2] Ibid. xxi. i. 1524. [3] Ibid. 855.

was drawn in the spring of 1546.[1] The inscriptions of 'Bollen' and 'Old man' are later additions and may be presumed to have been added by an English hand. Even if we cannot agree with William Paget that this French plat of the Boulonnais is the fairest and best that we have ever seen, it certainly is an unusual drawing from the point of view of drafting technique; we shall later find it useful for comparison with the drafting style which Rogers used on his own drawings of the Boulogne area.

The last drawing to be noticed is 'The Trew Conterfaicte of the State of the Pere' which is a view of the Dunette as it appeared on 26 January 1549—at least according to the inscription on the drawing itself.[2] It is fairly certain that John Rogers was not the draughtsman of this plat, because the drawing style, the elaborate use of colour, and the handwriting of the cartouche (indeed, the very use of a cartouche) are so much at variance with the other drawings ascribable to him. I have found not the slightest bit of documentary evidence which would connect him with the drawing; but then neither have I found any evidence as to who it was that did execute this interesting plat.

Besides the question of boundaries, there were other clauses in the Treaty of Camp which affected the development of English fortifications at Boulogne. The French agreed to pay to the English Government, in addition to sizeable pensions for Henry and his successors, two million crowns in gold to cover pensions in arrears as well as English expenses in the war and in the fortification of Boulogne. It was further stipulated that, if this sum had been paid within fifteen days after Michaelmas 1554, Boulogne and all the castles, towns, lands, rivers, and ports in the county of Boulogne would be returned in good condition to the French king. Finally, it was agreed that neither side would begin any new fortifications during the eight years that the English were to occupy the territory.[3]

Apart from the questions of whether the French really intended to pay that large indemnity or whether Henry really intended to relinquish Boulogne in 1554,[4] it was this last clause regarding fortifications which almost immediately created military and diplomatic difficulties. Both

[1] Curiously enough, the Citadel is not shown, although Clement's Bulwark is. On the French side, the drawing of Fort Châtillon is a later addition, and has in fact been pasted over the representation of the earlier ramparts.

[2] See above, p. 72.

[3] Rymer, op. cit. xv. 96–97; cf. *L. & P.* xxi. i. 1014.

[4] Cf. the astute remarks to this effect made by Odet de Selve to the Lord Protector in November 1547: *Correspondance politique de Odet de Selve*, ed. G. Lefèvre-Pontalis (Paris, 1888), 238.

sides were suspicious that the other would break this part of the treaty and gain some defensive advantage through new fortifications. This suspiciousness was alone sufficient to exacerbate relations between the two powers glaring at each other across the Liane River. But the problem was even more vicious in that both sides had been feverishly fortifying various points until the very moment when the treaty was signed. Who then could say whether this or that work was begun *before* or *after* 7 June 1546?

On the French side the problem mainly involved three sites where fortifications were being built—at St. Étienne, at the seaside village of Le Portel, and on the hill at the mouth of the haven (referred to later as Fort Châtillon). It was these latter two works which greatly agitated the English, for it seemed that French forts in these places would endanger the safety of the entire English port. Lord Grey sent Sir Thomas Palmer to report to the king on these works. Henry replied in a letter that nothing could be done about the matter, but then he privately instructed Palmer to tell Grey to destroy the works. Grey and the Council of Boulogne were puzzled by the conflicting commands, but they decided to follow the latter one.[1] On the night of 5 September 1546 a band of soldiers and five hundred pioneers crossed the haven and quietly overthrew the ditches and ramparts of the new works.[2] When Palmer was sent to England with the news that the mission was accomplished, Henry was overjoyed. But two weeks later, as the diplomatic repercussions of the English action were being felt, Henry cynically replied that Lord Grey had undertaken the enterprise without the king's command. He further reminded the French officials that sometimes officers, especially 'frontierers', will create military and diplomatic difficulties that might really become serious if it were not for the wisdom of their masters.[3] The French of course maintained that the work at Portel had been begun before the Treaty of Camp, so that they had the right to continue this project.[4] Nevertheless, in a draft treaty which was never formalized, they did agree in October to leave the works at Portel and Châtillon as they were at that stage, without completion or further additions.[5] St. Étienne was another matter. Although the English complained about the works proceeding in this town during the summer of 1546, there was little doubt that this fort had been begun in the preceding

[1] *A Commentary of the Services and Charges of William Lord Grey of Wilton*, ed. P. de M. G. Egerton. Camden Society, no. 40 (London, 1847), 5–7.

[2] *L. & P.* xxi. ii. 30.

[3] Ibid. 149.

[4] Ibid. 77.

[5] Ibid. 330.

spring. Thus in the draft treaty of October it was agreed that the French could continue their fortifications at St. Étienne. In this same treaty the English were allowed to continue the fortification of Ambleteuse, Blackness, and Boulogneberg, all of which, it was finally agreed, had been started before 7 June 1546.

By the winter of 1546/7 the major controversies centring around Boulogne had been settled, so that both sides could peacefully go about preparing for the next war. However, the English created a new series of incidents early in the spring of 1547 with preparations for and construction of the Dunette. We have already seen the repercussions of this undertaking, with the French claiming that it constituted a new fortification, while the English were insisting that it was merely a civil engineering project for the improvement of the harbour. In spite of opposition the English persisted in building the Dunette and provided thereby justification to the French for resuming old works or undertaking new ones as counter-measures. In May 1548 the French began again to fortify the hill at the mouth of the haven, opposite the Old Man.[1] Once again the English commanders at Boulogne were disturbed at the prospect of having a French fort overlook the mouth of the haven. The Lord Protector protested vehemently to Odet de Selve that this move was contrary to the Treaty of Camp, but the French ambassador merely raised the equally nagging question about the legality of the Dunette.[2] Deadlocked over the matter, and in spite of mutual warnings and threats as to the consequences thereof, both sides pressed forward the controversial projects.

Meanwhile, diplomatic relations deteriorated further because of increasing conflicts between the opposing forces garrisoned on either side of the Liane. In December 1547 the Protector protested to Selve about a recent French incursion into the Boulonnais in which several people were killed. During the next several months these skirmishes occurred periodically and created several momentary crises which had to be resolved through diplomatic channels. By the summer of 1548, although war was undeclared, English and French artillery were taking pot-shots at one another's fortresses across the haven.[3] It will be remembered that at this time Rogers was busily improving the defences of the Old Man—which had suffered somewhat from French cannonading—and that in particular

[1] *Cal. S.P. For., 1547–53*, 348–9.
[2] Selve, op. cit. 370, 406, 431.
[3] *Cal. S.P. For., 1547–53*, 353.

he was trying to provide a gate to the fort which would be less exposed to enemy fire.

It was obvious to all that the two countries were on the verge of war. Yet the English Government was in no financial position to get involved in another expensive war, and it was making every effort to avoid open and declared conflict. On the other side Henry II had little to lose and much to gain by pressing the issues at Boulogne. As early as March 1547 it was reported to the Council that the French king was preparing to retake Boulogne.[1] The Lord Protector, hoping to make the best of a bad situation, offered in November of that year to return Boulogne to Henry, if the English could retain Ambleteuse and the territory from the Slack River to the Calais Pale, and if the French king would support the marriage of Edward VI to Mary, Queen of Scotland.[2] Even though this meant that Henry would regain Boulogne seven years before the date set in the Treaty of Camp—and at far less expense to his Government than that treaty required—he rejected the offer, for he was confident that he would soon have Boulogne anyway, and perhaps Calais as well.

It was not until the summer of 1549, however, that Henry was prepared to make his move. By that time he had secured his defensive system across the Liane with forts at Châtillon, d'Outreau, and St. Étienne and had prepared his offensive weapon with the build-up of armed forces near Boulogne. On 25 August he swept into the Boulonnais at the head of his own army and assailed the fort at Sellaque with such vigour that within four hours the garrison of four to five hundred men capitulated. The same day he captured the town of Ambleteuse, whose inhabitants fled to the nearby fort after some eight of them had been treacherously killed by French soldiers while negotiations were taking place between the English and French commanders. Henry turned his guns upon the fort the next day, and by nightfall the twelve hundred men inside were prepared to make a conditional surrender: they would abandon the fort and all its artillery and munitions if Henry would let them go free with their personal belongings. Henry agreed, but remarked sarcastically that his mercy was greater than the English were accustomed to show the French. When the king turned upon Blackness, a mere blast from his bugler was sufficient to bring the four hundred soldiers to their knees, even if it did not bring the walls tumbling down. Henry was not satiated with the recapture of

[1] Ibid. 14. [2] Selve, op. cit. 237–8.

these three outlying forts; his next stop was Boulogneberg. To his surprise, when he arrived there on 30 August, the English garrison had already abandoned the fort, waiting this time neither for the boom of Henry's guns nor for the sound of his bugle. As they departed the English soldiers set fire to the fort, but Henry's own men arrived in time to extinguish the flames and save some of the supplies.[1]

Thus within a few short days all of the outlying fortresses in the Boulonnais had fallen into the hands of the French, and the English money and effort which had been expended on these works were swept away. Was the rapid capitulation of the garrisons the fault of inadequate fortifications and therefore to some extent at least the fault of John Rogers? This may have been the case at Ambleteuse. In 1564 two Italian engineers, Giovanni Portinari and Jacopo Aconcio, were summoned to Berwick-upon-Tweed for consultation on the fortification project there that was under the direction of Sir Richard Lee. Along with other conclusions and suggestions, they both agreed that the flankers of the bastions were too small for easy manœuvring of the guns within them. Portinari supported his contention by citing the unsuccessful defence of Ambleteuse when Henry II had besieged it in 1549. Henry's guns had smashed some of the English artillery inside one of the flankers, and because there was insufficient space in the flanker for repairing the pieces or replacing them, the flanker had to be abandoned. Since, Portinari argued, the flankers at Ambleteuse were identical in size and plan to those at Berwick, a similar situation might occur if the Berwick flankers were not enlarged.[2] If this later testimony of the Italian engineers was a fair judgement, and it probably was, it certainly indicates a miscalculation on the part of Rogers and his colleagues in planning the fort at Ambleteuse. Yet Rogers's responsibility for the rapid capitulation of Ambleteuse and the other outlying forts should not

[1] The preceding account of the fall of Sellaque, Ambleteuse, Blackness, and Boulogneberg was based on contemporary French documents: (1) letters from Henry II describing his capture of the forts (A. de Rosny, 'Documents inédits ou rarissimes, concernant les sièges de Boulogne, 1544–1549', *M.S.A.A.B.* xxvii (1912), 456–9); (2) letters from eyewitnesses who had participated in the enterprise (E. T. Hamy, 'Extraits de la correspondance de Maugiron relatifs à l'expulsion des Anglais du Boulonnais', *B.S.A.A.B.* vii (1904–7), 618 ff.); and (3) an anonymous account of the 'Voyage du roi Henry II en Boulonois, siège et prise des forts, chasteaux et ville d'Ambleteuse par l'armée royale l'an 1549' (ed. H. Gerard, 'Ambleteuse; un procès où se mêle l'histoire', *Annales Boulogne-sur-Mer*, ii (1904), 189 ff.).

[2] Iain MacIvor, 'The Elizabethan Fortifications of Berwick-upon-Tweed', *Antiquaries Journal*, xlv (1965), 86, citing the reports of Portinari and Aconcio, P.R.O., S.P. 59/8, ff. 110–17 and 125–7.

be pressed too far, because there were other and more important factors. The Privy Council was stunned by the news of the capitulations, and particularly the one at Boulogneberg, where the fort had been abandoned even before the French arrived on the scene. On 10 September the Council informed Lord Clinton at Boulogne that all the officers who had acted so ignobly were to be discharged from their duties and were to appear in person before the Council.[1] There was smell of treason, as the Council had noted when first hearing of the fall of Ambleteuse.[2] This suspicion was not without foundation, for the report of a French spy shows that there were sympathizers with the French cause, not only in Boulogneberg, but in the Old Man and in Boulogne itself.[3] These included natives of the Boulonnais in the service of England, as well as a couple of Englishmen (unnamed in the report) who were willing to betray the forts into the hands of the French.

In spite of Henry's initial rapid successes and the presence of would-be traitors in Boulogne, the French king was unable to complete his reconquest of the Boulonnais immediately. This was partly a result of his own indecisiveness—as Héliot has pointed out—in failing to capture the Old Man as he had the other outlying forts.[4] If he had taken this step, which he probably could have done without undue difficulty, the English position would have been untenable. Now that overland communications with Calais were severed, the garrisons at Boulogne were utterly dependent upon English shipping for their supplies. By capturing the Old Man Henry would have had both sides of the entrance to the haven in his hands, and it would have been virtually impossible for English ships to get in or out. But the Old Man remained in English hands, and the blockade which the French established proved to be only partly successful and of long duration. Throughout the autumn and winter of 1549/50 the English held on grimly at Boulogne, in spite of the lack of supplies and support from home, where the Protectorate was, to be sure, in its last throes. Something can perhaps be said for the effectiveness of the English fortifications at Boulogne in holding off the French besiegers indefinitely, whereas the earlier French works had fallen to the English in a matter of six weeks.

By January Henry was himself growing weary of the war and was

[1] *Cal. S.P. For., 1547–53*, 354. [2] Ibid. 46.

[3] D. Haigneré, 'Rapport d'un espion français en 1549', *B.S.A.A.B.* iv (1885–90), 275–8.

[4] Héliot, op. cit. 37.

willing to obtain Boulogne by negotiation, if not by conquest. Since the English Government was in no position to press the defence of the town, negotiations were begun which resulted in the Treaty of Boulogne, signed in March 1550. Therein the English agreed to relinquish Boulogne immediately, rather than in 1554. For their expenses in fortifying the town they would receive the sum of 400,000 gold crowns, instead of the 2,000,000 promised in the Treaty of Camp.[1] Thus on a minor key came to an end the 'Enterprise of Boulogne'.

The story of John Rogers, like that of Boulogne itself, is obscure during the later period of the occupation because of the increasingly sparse contemporary documents. In August of 1548 he was occupied in building a new gate for the Old Man.[2] Although he is not specifically named, he was most likely the one being urged in September 1549 to complete the small castle within the Old Man.[3] Thus he was still at work after the fall of the outlying forts and while Boulogne itself was under siege from Henry II. In February of 1550 a letter was addressed to him at Boulogne where he was directed as surveyor of works to settle a matter of wages due to a carpenter who had formerly worked at Blackness.[4] This suggests that he remained at Boulogne during the siege and continued to perform his duties until the very end of the English occupation.

[1] Rymer, op. cit. xv. 213.
[2] *Cal. S.P. For., 1547–53*, 353.
[3] Ibid. 355.
[4] Dasent, *A.P.C.*, ii. 396.

VI

THE LAST YEARS

THE last ten years of Rogers's life can be traced only sketchily, for though his general whereabouts are known, there is nothing like the detailed information that is available for the period of his career when he was surveyor of works at Boulogne. This dearth of evidence in the state papers reflects the fact that Rogers did not enjoy the same esteem in the eyes of Edward VI, Mary, and their ministers that Henry VIII had accorded to his surveyor. Rogers cannot exactly be said to have fallen from the grace of the monarchy; it was simply that his talents were not fully utilized by the royal will in those multiple duties which Henry had felt free to impose upon one of his favourite engineers. Nevertheless, the ever-varying life of those associated with building construction was not to be entirely denied to Rogers during his last decade. Even while he was still involved in the military projects at Boulogne, his growing reputation as a harbour engineer earned commissions for him to investigate problems connected with several harbours along the southern coast of England.

In June 1548 Rogers was instructed by the Privy Council to leave the works at Boulogne temporarily and go to Folkestone, Kent, where the citizens were desirous of obtaining a new mole to improve their harbour. Rogers was to meet certain other officials at Folkestone; together they were to decide upon the best means of constructing the mole and were also to estimate the cost of the project. The surveyor was then to draft a plat which would set forth their proposals.[1] This plat has apparently not survived, while the project itself does not seem to have moved beyond the planning stage. In these same instructions Rogers was urged to stop by Dover and view the conditions of the pier, which was made of timber and iron and was constantly in need of repair. A stone jetty might remedy this situation, the Council suggested, and Rogers should send a plat of Dover which would express his opinion on the matter. If Rogers did produce such

[1] Dasent, *A.P.C.* ii. 555.

a plat, it is not amongst the several sixteenth-century drawings of Dover harbour which are still extant.[1]

During the summer that Rogers was commissioned to make his surveys and plats of the harbours at Folkestone and Dover, he found himself called away from Boulogne to another civil engineering project on the coast of Kent, this time at Sandwich. The harbour at Sandwich, one of the important Cinque Ports of the southern coast of England, had long been silting up.[2] Since the latter half of the fifteenth century the townspeople had watched with growing concern the ruination of their haven, and thereby the prospect of economic ruin for their town. The silting of the haven was being produced by both natural and 'unnatural' factors. Sandwich itself lay inland; access to the sea was by means of the River Stour, which itself constituted the haven of the port. But the meandering Stour and its smaller tributaries were not producing a sufficient volume of water to keep the channel clear for shipping. Furthermore, the peculiar topographical feature of the area, which allowed tides coming from the north and east around the Isle of Thanet to meet in the Stour near Sandwich, did nothing to improve the deteriorating condition of the haven, and instead probably aggravated the situation. (See Pl. 24.) The flow of the river had been further impaired by man-made obstacles: sunken ships in the channel, weirs and kiddles placed in the stream for catching fish, garbage and waste thrown into the river. These conditions could at least be corrected by the townspeople; in due time the ships were removed, orders were given for the elimination of the weirs and kiddles, and prohibitions were issued against dumping rubbish into the river. But the natural causes of the town's dilemma could not so easily be met. During the reign of Henry VII there was a search for additional sources of water to feed into the River Stour for the purpose of scouring the haven. These efforts were to little avail, and the condition of the haven progressively worsened. Henry VIII interested himself in the problem and visited Sandwich in 1532 and 1539 in order to observe personally the difficulties facing the port. But nothing consequential resulted from these visits, so that by the opening of Edward VI's reign

[1] For reproductions and discussion of these drawings, see William Minet, 'Some Unpublished Plans of Dover Harbour', *Archaeologia*, lxxii (1922), 185–224; and Alec MacDonald, 'Plans of Dover Harbour in the Sixteenth Century', *Archaeologia Cantiana*, xlix (1938), 108–26.

[2] For the following, cf. William Boys, *Collections for an History of Sandwich in Kent* (Canterbury, 1792), 732 ff.; and Dorothy Gardiner, *Historic Haven: The Story of Sandwich* (Derby, 1958), 195–202.

the townspeople felt themselves to be in an extremely unfortunate position.

The records of the town show that in January 1548 the mayor and council were hoping that the national Government would come to the aid of the town; in May money was granted by the town council to certain local officials who were at Court seeking this aid.[1] Meanwhile, the mayor organized a survey of the haven to gather further information on the problem. It has been noted that he was assisted in this task by John Rogers, referred to at the time as 'Mr Surveyor'.[2] The conclusion which this study produced was that both Northmouth and the mouth of the Stour were too heavily silted up to be corrected by dredging or scouring the channel of the haven. Instead, it was decided that a 'new cut' should be made direct from Sandwich to the sea, which would avoid the problem of the silted-up Stour channel by simply by-passing it. If Rogers was in fact a member of the surveying team, then his influence can readily be seen in this proposed solution—it will be remembered that he had earlier suggested a cut at Blackness that would have turned the place into an island by extending the cut from 'sea to sea'. The conclusion regarding the haven at Sandwich was presented to the Lord Protector in a petition asking him 'to authorize the said mayor, jurats and inhabitants of the said towne . . . to cut out, newe erect, and make one newe cutt into their said haven in such place as . . . shall apere to be mooste apte, propise, commodious, necessary and mete for that same purpose'.[3] In this petition reference was made to a plat sent to the Protector to illustrate the condition of the haven as well as the proposed solution to the problem.

The Government responded to the petition on 4 August 1548 by appointing a commission to investigate the proposal for new harbour works at Sandwich. Besides the Archbishop of Canterbury and four other men, the commission included Sir Richard Lee. Curiously enough, the letter of appointment was also addressed to Edward Thwaytes, Robert Legg, and John Rogers, although these three were not actually named to the commission.[4] There can be little doubt, however, that Rogers was the chief

[1] Maidstone, Kent Archives Office, SA/AC3 (Sandwich Old Red Book, 1527–51), ff. 204 and 207^{v}. Presumably the latter is the 'Old Red Book, 1527–51', although I was unable to find a reference to the survey in this massive volume.

[2] Gardiner, op. cit. 202, provides these details about the survey, but does not mention the exact date or the source of her information.

[3] Petition quoted in Boys, op. cit. 734.

[4] Ibid. 734–5.

technical official delegated by the Government to the project, for in the report of a similar commission appointed in 1559 there was mention of 'the newe cute at Sandwiche lately begunne by devise of John Rogers appoyncted therfore in kinge Edwardes the vith tyme'.[1]

On 9 November 1548 the Privy Council granted one shilling to 'John Rogers' servant of Bulloyn, resorting to the Court with the plat of Sandwych'.[2] It may be safely assumed that this plat had been drafted by Rogers himself, and it probably was the product of another visit to Sandwich which he made after receiving the appointment as the chief technical official for dealing with the harbour problem. Since this plat does not seem to have survived, we may use one from the Elizabethan period which illustrates the problems faced by Rogers and the citizens of Sandwich (Pl. 24). Sandwich is located on the south end of a loop in the Stour, where the river flows southward and then turns northward, with the sides of the loop running roughly parallel to the seashore. The plan which Rogers devised was to make a cut from the south-eastern corner of the loop to the sea, thus enabling ships to sail directly from Sandwich Bay into the haven without having to navigate the tortuous and silted channels of Pegwell Bay and the lower Stour. However, work did not begin immediately on the project, although Rogers made occasional trips to Sandwich to study further the problem of improving the haven. In March 1549 the town fathers decided 'that the charges of the said Surveyor from time to time shall be borne by the town and that the Treasurer for this present year being shall provide all victuals as shall be meet for the table of the said Surveyor, wheresoever it shall please him to be, for and in recompence of his travail and pains taken in riding and viewing the said haven'.[3]

By the middle of April work was under way.[4] Men were set to digging at a point about fifteen hundred yards downstream from Sandwich, where the cut left the river and headed in a south-easterly direction for the sea. Skirting the rough pasture along the coast, the cut when completed would emerge at the sea in the low land just south of Sand Hills, between the two areas now occupied by the Royal St. George's and the Royal Cinque Ports Golf Links. (See Pl. 24.) Progress was slow, however, because from the beginning the project was plagued by financial troubles. The expected

[1] Boys, op. cit., 736.

[2] Dasent, *A.P.C.* ii. 228.

[3] Old Red Book, 1527–51, f. 215^{v}.

[4] Ibid. f. 217.

income from the king and other sources was not forthcoming, so that money had to be borrowed from certain citizens of Sandwich for paying the workmen.[1] Through the working seasons of 1549 and 1550 the project was kept going only by such *ad hoc* arrangements. In March 1551 urgent petition was made to the Government for redress of these grievances, and again in August.[2] By November it was admitted that because of lack of financial support, work on the project had ceased.[3] It was not revived again in Rogers's time.

But a portion of the cut was there to remind later commissions of Rogers's scheme for solving Sandwich's problem.[4] Indeed, for the next two centuries the various proposals to save the harbour followed the scheme put forward by Rogers. In the surveys of 1559, 1573, and 1737 Rogers's plan was generally accepted, although it was felt by some of the commissions that he had placed his 'new cut' too far downstream, and that a more southerly cut would be less costly because it would pass through land four feet lower than that traversed by Rogers's cut.[5] Nevertheless, these later proposals came no nearer fruition than had Rogers's, so that gradually the ruination of Sandwich haven brought an end to the town's long history as a seaport.[6]

With the cessation of work at Sandwich, Rogers apparently retired to Folkestone for the winter season. Perhaps he had become interested again in constructing the mole at Folkestone which had been projected two years earlier. In January 1551 'Mr. Rogers of Folkestone' was instructed to communicate with Sir Thomas Cheyney, Treasurer of the King's Household, regarding 'the situacion and making of a Bulwerke at Sheres Neyshe'.[7] This was the fort of Sheerness, located on the Isle of Sheppey on

[1] Ibid. f. 219.

[2] Ibid. ff. 243, 247v.

[3] Ibid. f. 251.

[4] And, in fact, it is still there in the lowlands to the east of Sandwich. See the one-inch Ordnance Survey Map (1959), Sheet 173, where the 'New Cut' and 'Old Haven' are clearly indicated.

[5] This was the conclusion of a Dutch engineer, Andrian Andrison, in 1573. For a copy of his report, see Boys, op. cit. 736 ff. Cuts similar to the one projected by Andrison were proposed on later drawings. See Pl. 24 and A. H. W. Robinson, *Marine Cartography in Britain* (Leicester, 1962), Pl. 29, which reproduces the drawing made by Charles Labelye in 1737 (Admiralty, Hydrographic Department, A 38 Dc).

[6] Of the Cinque Ports (Dover, Sandwich, Romney, Hythe, and Hastings) only Dover has survived as a functioning port today. In the sixteenth century all five were plagued with the physical deterioration of their harbours caused by beach drifting and/or silting up of havens. See James Bird, *The Major Seaports of the United Kingdom* (London, 1963), 240–1; and especially Montagu Burrows, *Cinque Ports*, 2nd ed. (London, 1888), 1–21, for descriptions of these natural processes.

[7] Dasent, *A.P.C.* iii. 193.

a promontory where the River Medway enters the Thames estuary.[1] If this Mr. Rogers was the military engineer, he was apparently being called upon merely for technical advice, for there is no indication that John Rogers assumed a continuing appointment at Sheerness. In the same way he seems to have spent some time during 1550 at Guernsey in the Channel Islands, for in March 1551 he was granted £40 to reimburse himself 'in his chardges going this last year to Garnesey for the Kinges Majesties affayres'.[2] The captain of Guernsey, Sir Peter Mewtis, was paid £492. 17*s.* 10*d.* in 1552 for repairs and fortifications on the king's castle of Guernsey, that is, Castle Cornet.[3] Professor John Le Patourel has suggested that this work produced the bastion known as 'Mewtis Bulwark', and that Rogers's expenses at Guernsey might have been connected with this project.[4] This is certainly a plausible hypothesis. It would be questionable, however, to press it further and suppose that Rogers was responsible for the design of Mewtis Bulwark, since the round form of this bastion would have been quite old-fashioned for the engineer who had spent six years on the technically advanced projects at Boulogne.[5] Probably Rogers was sent to Guernsey merely in a consultative or administrative capacity, rather than as the designer or supervisor of construction.

It appears that Rogers, having completed his tour of duty at Boulogne, was being utilized by the Government for various odd jobs and consultations, until some major project might develop where his talents could be put to fuller service. This opportunity seemed to present itself in the winter of 1551 when the Government decided to fortify a number of places along the southern coast of Ireland. During the autumn of 1550 the Lord Deputy of Ireland, Sir Anthony St. Leger, had surveyed the forts in Leix and Offaly and found them sadly wanting in discipline and defence. The fortifications along the southern coast of Ireland were likewise in unsatisfactory condition, particularly since there was suspicion of a French invasion of the island along this coast.[6] In February 1551 Sir James Croft was sent to Ireland with instructions to 'view and consider' the towns,

[1] For a 'Prospect of Sheerness' see the chart of this area of the estuary published by Captain John Mitchell in 1733 and reproduced in Robinson, op. cit., Pl. 25.

[2] Dasent, *A.P.C.* iii. 227.

[3] Ibid. iv. 69.

[4] John Le Patourel (ed.), *The Building of Castle Cornet, Guernsey*. I, *Documents Relating to the Tudor Reconstruction* [Guernsey], 1958, pp. 10–12.

[5] See the plan and views of Mewtis Bulwark ibid., Figs. 1, 4, and 5.

[6] Richard Bagwell, *Ireland under the Tudors* (London, 1963), i. 351.

castles, and havens along the coast—including Waterford, Youghal, Cork, Kinsale, Baltimore, and Berehaven, as well as points in between—for the purpose of determining the works necessary to make these places defensible.[1] He was to survey older military establishments for their usefulness and choose new sites to be fortified when these seemed necessary. In order to inform the Government of his conclusions he was to send plats of the areas surveyed and the works proposed. This was indeed a grand scheme, and if it had been carried through with the completeness implied in the instructions, the English Government would have found itself sponsoring a series of projects that would have equalled or exceeded Henry VIII's string of forts along the southern coast of England. But of course the Government of Edward VI had not the ability to pay for such a scheme, even if it had the intention of doing so.

Nevertheless, Croft set about his task diligently, and in his survey of the fortifications he was aided by John Rogers. Both of these men had been commissioned to go to Ireland about the same time. On 14 February Rogers received a grant of £600 in prest money, 'to be by him ymployed for the provision of showels and spades, mattockes and other necessaries for Irelande'.[2] In April he was listed with seven other men who had received £11,544. 19*s*. 0*d*. from the High Treasury during the period 1 October 1550 to 25 April 1551.[3] Presumably this sum had been directed towards the expenses of the garrisons and works in Ireland. Meanwhile, Rogers had been engaged with Croft in making the survey. On 11 May the Privy Council wrote to Croft (who had by then become Lord Deputy of Ireland, replacing St. Leger) that it had received the plats of Cork, Kinsale, and Baltimore, 'sent by Rogers's man'.[3] Having by now seen how often Rogers was called upon to make plats, and noting the familiarity of this procedure in which the plats were brought to the Council by Rogers's own servant, we may assume that these plats also were drafted by him. Unfortunately, none of them seems to be extant.[4]

The precise nature of Rogers's official position in Ireland is uncertain. Sometime after 11 May Croft responded to the letter from the Privy

[1] *Cal. Carew MSS., 1515–74*, 230–1.

[2] Dasent, *A.P.C.* iii. 213.

[3] *Cal. S.P. Ireland, 1509–73*, 112.

[4] A very roughly sketched plat of Cork county in the P.R.O. (S.P. 63/30, no. 106) is definitely not by Rogers; it dates rather from the Elizabethan period. Nor does Rogers appear to have drafted any of the other maps of Ireland that are listed in the 'Index of Maps and Plans in the Public Record Office, *c.* 1430–*c.* 1603' (P.R.O., Round Room, Press R 7).

Council by asking that Rogers 'be employed for furtherance of the fortifications'.[1] This implies that, even though Rogers was cited in the financial document of 25 April 1551, he had not yet been appointed to an official position, such as surveyor of works.[2] Whether or not he ever received such an appointment remains open to question, for his name disappears from the correspondence relating to Ireland. Probably the reason he does not figure more importantly in the proposed fortification works is that those works never really got under way at this time. The reason was the usual one—lack of funds to pay the workers. In May the Government informed Croft that it was sending soldiers and ships with large supplies of food and other things, and that the men were to be employed during the summer in the fortification works at the various places previously determined.[3] On 20 May Croft replied that 1,000 soldiers and 120 pioneers had indeed arrived from Bristol, but that he was greatly distressed because there was no money with which to pay them.[4] This financial plight doomed the building plans from the beginning, and in spite of the Government's desire that the local towns should share the expense of fortification, there is little evidence that any significant works were accomplished at this time.

Now that the great plans for fortifications in Ireland had collapsed and Rogers's services were no longer needed there, he fell back into the position of sometime technical consultant and handyman. In February 1552 he and Sir Richard Cotton, who had been the treasurer of Boulogne, were instructed to inspect the condition of Guines Castle and 'to make out a new plat of Guysnes in suche perfect sort as at their cummyng hither they may substancyally declare the state of the thing from point to point'.[5] It would be pleasing to record that this drawing has survived and is 'The Plot of Guins' (Pl. 5) that was used earlier in analysing the Guines Castle drawings produced by Rogers in 1541. In 1552 the castle perhaps appeared at the stage of development that is shown on this drawing. But the drafting technique of the plat—with the peculiar projection of the castle walls and the vibrant use of colour—is so at variance with Rogers's customary style that we can hardly attribute it to him. Perhaps for

[1] *Cal. S.P. Ireland, 1509–73*, 114.

[2] At this time there was a surveyor of mines and a surveyor of lands in Ireland (*Cal. Carew MSS., 1515–74*, 227–8), but I have found no reference to a surveyor of works.

[3] Ibid. 231.

[4] *Cal. S.P. Ireland, 1509–73*, 112.

[5] Dasent, *A.P.C.* iii. 475.

some reason Cotton alone produced this plat, though admittedly this is merely a guess.

On 22 February Rogers and Cotton received further instructions from the Privy Council. After their survey of Guines they were to return to England by way of Dover, where they were to inspect the old pier, which had continued to decay since Rogers had surveyed it in 1548. Rogers was to produce a plat to show the condition of the pier.[1] Twenty months later, in October 1553, Rogers was again sent to inspect the pier, this time in the company of Sir Richard Lee and Benjamin Gonson. They were to see whether it could be sufficiently repaired for use, and how much it would cost. Furthermore, they should consider any other devices which might be constructed to improve the haven.[2] It is known that Dover Harbour was facing some of the same difficulties in silting and beach drift that were ruining the haven at Sandwich. Indeed, during the preceding fifty years a number of proposals and attempts had been made to correct the problems.[3] If the commission of Lee, Rogers, and Gonson presented a new scheme at this time, nothing seems to have resulted from it.

Meanwhile, a new opportunity had seemingly opened for Rogers in May 1552 when he was ordered to Portsmouth 'to vieu and peruse the state of thinges there, and to devise what wayes may best be taken for the fortefying of the same'.[4] Shortly thereafter he was granted £40 for his expenses in going to Portsmouth.[5] Yet the precise character of his office eludes us. There is no indication that he was made surveyor of the works at Portsmouth, a position that might have given him the security enjoyed by Richard Popinjay, one of the foremost engineers of the next generation, who held the office for a quarter of a century. Instead, it appears that Rogers went in that capacity which he was accustomed to fill during the later years of Edward's reign—that of technical consultant without official position. So the opportunity at Portsmouth never fully materialized, and Rogers found his services being less and less utilized by the monarchy. Particularly after Mary came to the throne, engineers like Rogers and Lee, who had played conspicuous roles in military and even political affairs during the time of Henry VIII, faded into the background.

In April of 1554 Rogers and Robert Broke, clerk, received £100 to be

[1] Ibid. iii. 484.

[2] Ibid. iv. 356.

[3] On these, see John Lyon, *The History of the Town and Port of Dover and of Dover Castle* (Dover, 1813), i. 150 ff.; and Minet, op. cit. 185 ff.

[4] Dasent, *A.P.C.* iv. 47.

[5] Ibid. 56, 67.

employed in and about certain of the queen's affairs, but what they were doing remains undetermined.[1] Except for this incident and the commission appointed in the autumn of 1552 to inspect Dover pier and haven, John Rogers disappears from the public record from the spring of 1552 until the autumn of 1556. This does not necessarily mean that he performed no services for the monarchy during this period, but if he did they have gone unrecorded. Then in the autumn of 1556 things began to move for Rogers. On the third day of October he was summoned to St. James's Palace for some unknown purpose; a month later he was appointed surveyor of the queen's works at Berwick.[2] Scattered efforts had been made during the reigns of Henry VIII and Edward VI to transform the medieval town walls of Berwick into a defensive system more suitable for contemporary warfare, and now Mary's Government was trying its hand at the task.[3] Thomas Lord Wharton, Warden of the East and Middle Marches, had already received £1,000 for this project, and Mary, concerned that the work proceed 'with as much husbandry as may be', was now sending Rogers to confer with Wharton and to serve as surveyor of the works. The two men were to review the whole project, devise how best it might be forwarded, and decide whether part of it should be let out to contract. Furthermore, Wharton was to indicate to Rogers how much of the £1,000 was still available and what unused materials remained for the works.[4] One wonders how well Rogers got on with this great lord of the north, who was here being asked to render account to the newly-appointed surveyor of works. Rogers's stay in Berwick was a very brief one, although there is no evidence that this was the result of altercations between himself and Wharton.

In spite of his short tenure at Berwick—it was less than a month—Rogers seems to have had at least some effect on the planning of the fortifications, for in October of 1557 his work was reviewed to see if it should be continued as he had devised it.[5] Yet he probably did little more than make further *ad hoc* proposals of the type that had been characteristic of the work at Berwick during the two preceding reigns. It remained for Rogers's friend and colleague, Sir Richard Lee, to rethink the whole defensive system of Berwick. As surveyor of the works after January 1558,

[1] Dasent, *A.P.C.* v. 16.

[2] Ibid. vi. 3, 22.

[3] Iain MacIvor, 'The Elizabethan Fortifications of Berwick-upon-Tweed', *Antiquaries Journal*, xlv (1965), 64–65.

[4] Dasent, *A.P.C.* vi. 22.

[5] Ibid. 192.

Lee provided the design and supervised much of the construction of the extensive bastioned fortifications which survive today, and which constitute the most impressive extant monument of Elizabethan fortification.

For six years Rogers had been waiting for the kind of appointment which he finally received in November 1556 as surveyor of works at Berwick. Yet within a month he was removed from that office to become surveyor of the fortifications, works, and repairs within the town and marches of Calais.[1] This was an even more prominent position than the one at Berwick, and it was relatively well paid; he received £30 annually plus a house in Calais. Yet he became the surveyor at a most unpropitious moment, for at this time Philip of Spain was leading Mary, and the English nation with her, into war with France. In such a war the Calais Pale was again likely to be a focal point of offensive and defensive manœuvres on both sides. The defences at Calais, however, were altogether unsatisfactory from the English point of view. Although work had been carried on steadily for a quarter of a century, there had never been a major renovation of the late medieval fortifications to meet the requirements of modern artillery warfare. By the time Rogers became surveyor of works late in 1556 there was little hope that such a reworking of the defences could take place, for by this date Philip and Mary's Government was in no financial condition to support such a programme.

Nevertheless, some attention was given to the problem in May of 1557, when the Government outlined a fairly extensive programme of new works for Calais. Several new walls leading from the town to the harbour were planned, as well as three new bulwarks and a new gate.[2] But there is little indication that these works were ever executed. This must have been frustrating to Rogers, who had taken part in the planning of the works, and whose wide experience in fortifications would indicate to him that the Calais defences were in an unsatisfactory condition. But without the money there was little he could do about it.

As the surveyor of Calais Rogers was also responsible for the works in the outlying forts of the Calais Pale. It is ironic that the last position which he held placed him in charge of the works at Guines Castle, where he had first gained a prominent position in the royal service. The irony is heightened by the fact that William Lord Grey, his former superior officer and sometime antagonist at Boulogne, was the captain of Guines. A letter

[1] *C.P.R., 1555–7*, 527. [2] Dasent, *A.P.C.* vi. 91.

from the Privy Council in June 1557 suggests that Grey and Rogers were once again in opposition. Apparently Grey had written to the Council asking for authority to prescribe the works at Guines which Rogers should pursue. But the Council replied that it was the prerogative of the king and queen to decide such matters, 'uppon good deliberacion and with thadvise and consent of the Lord Deputie and Counsaill of Callaice'.[1] Whether or not this rebuff meant a victory for Rogers depended on how readily he could make his will felt with the Lord Deputy and Council of Calais, and this much we do not know. Probably the question was as academic then as it is now, for without sufficient funds from the Government it made little difference who 'devised' the works, since they could not be carried out anyway. It thus seems unlikely that during his tenure as surveyor Rogers effected any appreciable changes in the fortifications of Calais or the other forts within the Pale.

War between England and France erupted in the summer of 1557, and the forces were set in motion which, within a matter of months, resulted in the loss of Calais to the French. In spite of the ominous and obvious warnings in the latter part of 1557 that the French were laying plans for an attempt on Calais, the English Government either failed to recognize the warnings or else refused to give sufficient heed to them, for little was done to prepare the garrison or the fortifications against the imminent attack from the French.[2] Thus, when the French moved into the Pale on 1 January 1558, the outposts fell one by one—first Sandgate, then Ruisbank, and then Newnham Bridge. By the fourth day of January Calais itself was under siege, and on the seventh the town capitulated.[3] Now only Guines remained in the hands of the English. Thanks to the more adequate defences which Guines had acquired after 1540 and to the valiant leadership of William Lord Grey, the garrison held out through eight assaults until 20 January, when it capitulated on terms that allowed the soldiers to return to England fully armed.[4]

The part played by Rogers in the last days of the English occupation of the Calais Pale remains obscure, for there is only one tantalizing bit of

[1] Dasent, *A.P.C.* vi. 101–2.

[2] A. F. Pollard, *The History of England from the Accession of Edward VI to the Death of Elizabeth* (London, 1929), 167 ff.

[3] On the fall of Calais, see the collection of documents in *An English Garner: Tudor Tracts, 1532–1588*, introd. by A. F. Pollard (New York, 1903), 290–317.

[4] For the siege of Guines, see *A Commentary of the Services and Charges of William Lord Grey of Wilton*, ed. P. de M. G. Egerton. Camden Society, no. 40 (London, 1847), 18–39.

evidence. In a muster roll of the Calais military personnel dating from September 1557 a certain John Rogers is named as the leader of one hundred infantrymen.[1] Could it be that Rogers the surveyor had assumed one more role in his multi-coloured career and was now acting as a lesser officer in the army during the crisis produced by the war? This cannot be definitely asserted, but Rogers's long affiliation with military life as a specialist in fortifications would have given him sufficient experience to fill at least a minor post. Whether Rogers actively participated in the defence of Calais is debatable, but his fate at the end of the English occupation is more certain.

When Calais fell to the Duke of Guise on 7 January 1558 the small English garrison was allowed safe conduct out of the town. Only Thomas Lord Wentworth, Lord Deputy of Calais, and fifty of the military officers and officials of Calais and the outlying forts were retained as prisoners for ransom. Two weeks later when the French captured Guines the same conditions were imposed upon William Lord Grey and several of his officers.[2] It is fairly certain that John Rogers, surveyor of works and thereby a high-ranking civil official, was taken prisoner at the fall either of Calais or of Guines, and circumstances point to the former. He was still a prisoner on 16 March, when the Privy Council ordered the mayor of Rye in Sussex to permit 'William Flynte, servaunte to John Rogers, late Surveyour of Callays, to repayre to his saide maister, being prysoner in Fraunce, with such thinges as he hath to bring to his saide maister for his relyef'.[3] Thus Rogers's career as a military engineer brought him to the fate that always looms as a possibility for those directly involved in warfare—that of becoming a prisoner of war. But the price which he paid for his position at Calais was more than his freedom, for it seems likely that he died as a prisoner of the French.

One may well imagine from the wording of the letter to the mayor of Rye that Rogers was a sick or wounded man who needed 'the things for his relief' which his servant was bringing. He apparently died within a week or two after this letter was written on 16 March 1558. This conclusion is based upon the stipulations of a royal grant made two years later. In February 1560 William Painter was appointed Clerk of the Ordnance to

[1] Historical MSS. Commission, *The Manuscripts of the Right Honourable F. J. Saville Foljambe* (London, 1897), 6.

[2] *Tudor Tracts*, 293, 298; *William Lord Grey of Wilton*, 36–37.

[3] Dasent, *A.P.C.* vi. 285.

replace John Rogers, deceased, to whom the office had been granted for life by patent on 28 February 1554. Painter was to receive wages of 8*d.* a day, payable from Lady Day (i.e. 25 March) 1558.[1] The implication of this grant is that Rogers was dead by 25 March 1558, since it was to that date that Painter's wages were made retroactive upon his appointment as Clerk of the Ordnance. It was not unusual at this time for the recipient of an office to receive pay from the time of his predecessor's death rather than from the moment when he actually assumed the office. Rogers himself had only recently become surveyor of works at Calais on these terms.[2] If Rogers died in March 1558, as the wording of this grant implies, then he very likely died as a prisoner of war in France.

This conclusion regarding the date and circumstances of Rogers's death lends a certain poignancy to his story; at the same time it clarifies several biographical details. It shows that the John Rogers who was serving the queen in Ireland as late as 1572, or the John Rogers who was the mayor of Poole in that same year, was not John Rogers the military engineer, for the latter's career had come to an end fourteen years earlier with the fall of Calais.[3] It helps to explain the fact that in July 1559 an annuity of 20 marks (£13. 6*s.* 8*d.*) was granted for life to Agnes Rogers, wife of John Rogers, in memory of the latter's services to Henry VIII, Edward VI, and Mary.[4] The relatively large size of this annuity may be construed as recognition of the *special* services rendered to the English monarchy by Rogers in a career which culminated in his death as a prisoner of war in France. Finally, the date of his death in March 1558 supplies a fact missing from an important document in Rogers's story: his last will and testament. Rogers made his will on 6 August 1557 while he was surveyor of works in Calais; it was probated on 10 May 1559. The probate records do not give the date of his death, and since one cannot decide *a priori* how much time had elapsed between the death of the testator and the proving of the will, external evidence must be sought. This fortunately has been supplied by the information in the grant to William Painter. Just why fourteen months elapsed between the time of Rogers's death in March 1558 and the probate of his will is not clear, but it could have involved one or more of several factors: Rogers's death in a foreign country, his property holdings

[1] *C.P.R., 1558–60*, 281.

[2] Ibid. *1555–7*, 527.

[3] References to these other men may be found in *Cal. S.P. Ireland, 1509–73*, 479; and *Cal. S.P. Dom., 1547–80*, 451. They have been incorporated into the biographical sketch of Rogers the engineer by Robinson, op. cit. 148.

[4] *C.P.R., 1558–60*, 55.

in several counties, the inherent slowness of the probate court, or some other situation completely unknown to us.

Rogers's last will and testament is an interesting document, for it reflects the man's personality in spite of the set form and stereotyped phraseology of such instruments.

In the name of God, amen, the sixth day of August in the year of the incarnation of our Saviour Christ, a thousand five hundred fifty and seven: I, John Rogers, surveyor of the queen's majesty's town and works of Calais, being daily vexed with certain diseases in body but (praised be God!) perfect and whole in memory, do make this my last will and testament. First, I bequeath my soul unto my Lord God the Father Almighty, assuredly believing that through the death of his son, Jesus Christ my Saviour, to enjoy that heavenly kingdom which is prepared for his elect children; my body to be buried in the earth at the discretion of my friends, without any pomp or vain ceremonies. Secondarily, I bequeath unto Agnes Rogers, my wife, and her heirs all my lands and goods, moveable and unmoveable, wheresoever they be, or may be found, as well within the town of Calais as within the realm of England: as namely my house in London called Saint Giles Hall, with all the appurtenances thereunto belonging; as also my lease of the lordship of Streat in Somersetshire, with my lease of the priory of Folkestone in Kent, to have and to hold to her own proper use to bestow them according to her will—and being sorry that it is no more. Also I make my said wife my full and sole executrix, to receive and pay all that is my right dues and debts. In witness whereof I have made this my last will or testament and have thereunto set my name with my own hand, the day and year before written, in the presence of these witnesses whose names and hands are hereunder written. By me John Rogers; witnesses to this present, John Wynsted and the mark of William Swinbery, with other.[1]

Only a couple of points in this document need be commented upon. Perhaps too much emphasis should not be placed on the exact wording of the section concerned with religious matters, but it may be suggested that the passage does have a Protestant overtone, perhaps reflecting Rogers's own religious sentiments. If so, this might help to explain why Rogers's services were no longer required when Mary came to the throne, and why he had to wait for years before he again achieved an important position in the royal building works. Secondly, even though Rogers regrets that he has no more property to give his wife than that stated in the will, it should be noted that for a man who had risen from the ranks of working masons, his holdings were in fact considerable. He acquired one of these

[1] Maidstone, Kent Archives Office, Probate Records Canterbury Diocese, A. Vol. 32, f. 30.

properties as a result of the royal favour which he enjoyed under Henry VIII. On 17 January 1548 John Rogers, gentleman, was granted a messuage and house ('Le Comen Hall') in St. Giles without Cripplegate, London. This was the 'Saint Giles Hall' mentioned in Rogers's will. The grant explained that Henry VIII had promised this holding to Rogers, 'which certainly would have been confirmed to him by letters patent if there had been time before the said king's death'.[1] On the other hand, the manor of Streat which Rogers held in Somerset was obtained during the time and at the behest of the Government of Edward VI. In May 1552 orders were given 'to make a lease, for xxj yeres, of the mannour of Strete, parcell of the possessions of the late Monastery of Glastonbury in the countie of Somerset, to John Rogers, reserving thaccustomed rent and the woodes belonging to the same'.[2] The manner in which he obtained the lease on the priory of Folkestone has not been ascertained, but it will be remembered that he had been active in the Folkestone area around 1550—perhaps he got the lease at that time. Of the 'lands and goods, moveable and unmoveable . . . within the town of Calais', we may suppose that he had accumulated these since his appointment as surveyor of works there in December 1556. The terms of that appointment had included not only wages of £20 a year, plus a life annuity of £10, but also a messuage and mansion in the parish of St. Nicholas in Calais, at the same rent which Rogers's predecessor, William Pettyt, had paid.[3] From all of this it is clear that Rogers's career as a surveyor of works in the royal service had brought him not only a reputation as a military engineer and platmaker, but a fair amount of property as well.

[1] *C.P.R., 1547–8*, 294. The identity of this John Rogers, gentleman, is assured by a confirming grant of 25 December 1548, where the holder of the house in St. Giles is specifically referred to as John Rogers, surveyor of Boulogne: Dasent, *A.P.C.* ii. 233.

[2] Dasent, *A.P.C.* iv. 48.

[3] *C.P.R., 1555–7*, 527.

VII

JOHN ROGERS AS A MILITARY ENGINEER

One of the most intriguing and perplexing problems in the history of medieval and early modern building is what may be called the professionalization of architecture and engineering. The problem involves a number of related issues: who was responsible for design and supervision of building? what was the relation between the designer and the patron of the building? what educational background and special training did the designer normally enjoy? what was his social and economic status? and finally, when did engineering become a profession separate from architecture? It is not our purpose here to try to answer these questions in full, for they are intricate problems that would require lengthy discussion. Nevertheless, a consideration of these questions in the light of Rogers's career will bring out some interesting points regarding the place of the sixteenth century in that long development which eventually produced the professions of architecture and engineering.

Rogers's career occurred during that period of the sixteenth century which saw on the one hand the gradual demise of 'Gothic' architecture in England, and on the other, increasing attention being given to classical architectural forms that were inherited from ancient Rome through Renaissance Italy and France. Accompanying this change of taste in architectural style were changes in the status and functions of the architects and engineers who were bringing in the new style. It should be remembered that these changes did not occur overnight. Just as scattered samples of 'Gothic' buildings continued to be constructed in England even into the seventeenth century, so likewise was the transformation of the 'medieval' into the 'modern' architect a gradual process, one that was only completed early in the nineteenth century.[1] Nevertheless, this period in the sixteenth century was the beginning of the end for the medieval

[1] For surveys of this subject, see H. M. Colvin's introductory essay on 'The Architectural Profession' in his *Biographical Dictionary of English Architects, 1660–1840* (Cambridge, Mass., 1954), 10–25; and especially, Frank Jenkins, *Architect and Patron: A Survey of Professional Relations and Practice in England from the Sixteenth Century to the Present Day* (London, 1961).

craftsman/architect. Rogers was himself a part of these changes, and his career illustrates in many ways those forces that were to bring about this transformation. In order to understand his place in the changing role of the architect and engineer, some attention must be given to both the medieval and Renaissance background of his work.

In medieval English building almost invariably the architect and engineer had been trained originally in one of the building crafts. This meant that he received a thoroughly practical training in the art of building, one that made him quite familiar with the problems of construction as well as of design. Furthermore, his education *as an architect* involved little, if any, book-learning. He was taught directly by the masters of his craft, who passed on to him those precepts and practices which had been formulated from the experiences of past generations of building craftsmen.[1] The normal education and career of the medieval architect thus involved an apprenticeship in masonry or carpentry, a few years' work as stonecutter, stonesetter, or joiner, a period of service with a master builder as his warden (i.e. under-master), and finally, when sufficient experience had been acquired, a position as the master mason or master carpenter, responsible for the design of the building. It is in this sense that one speaks of the medieval craftsman/architect, and for this and the following reasons one can hardly speak of a profession of architecture in the Middle Ages. Architecture had no status as a learned discipline within the curriculum of the medieval university. What is equally significant, there were few learned men who wrote works dealing with architecture or architectural theory. Since the craftsmen themselves did not write books, the Middle Ages were almost totally devoid of treatises devoted to the theory and practice of building.[2]

If architecture as a subject of discussion for learned men did not attract much attention, architecture as a profession likewise did not attract men from the higher classes of society. The status of architecture is clearly

[1] For a more elaborate discussion of this subject, see my article, 'The Role of the Master Mason in Mediaeval English Building', *Speculum* xxxix (1964), 388 ff.

[2] In his monumental book, *The Gothic* (Princeton, 1960), Paul Frankl seeks every scrap of medieval evidence relating to the theory of Gothic architecture, but finally has to admit that 'it is astonishing that the scholastics, all of them clerics or monks, themselves witnesses of the growth of the cathedrals, who celebrated mass in them, heard confession, preached, who had a voice in the plans, superintended the building activity, managed the finances, who philosophized on the "beautiful" and wrote treatises, never refer to Gothic by a single word and do not mention a single cathedral' (p. 107).

indicated by the fact that throughout the Middle Ages the role of the architect continued to be filled by commoners who plied the building crafts. This is not to say that a medieval architect could not acquire status. It was quite possible for a master mason to accumulate a fortune and to attain social eminence and even local political prominence, as did Henry Yevele in the last half of the fourteenth century.[1] But the point to note is that even though men like Yevele might achieve prominence through their work as architects, still the status of architecture was such that it did not attract into its ranks the sons of nobles, gentry, or even prosperous townsmen. The highest positions in architecture were filled by commoners who had risen through the ranks of the building crafts, and by the sons of these same men, who likewise had to work their way up from apprenticeship to master mason or master carpenter.

During the fifteenth century this condition of architecture was considerably altered in Italy, where under the impact of classicism a new architectural style was adopted and a new view of the architect developed. With the growing desire to imitate the building forms of classical Rome, architecture became more and more a learned subject—bookish, antiquarian, and archaeological. For the first time in well over a thousand years learned men turned their attention to writing treatises solely devoted to architecture: witness the fifteenth-century works of Leon Battista Alberti, Francesco di Giorgio Martini, and Antonio Filarete. The result of this transformation of architecture into a subject for learned study was to produce a new kind of architect, one who combined the interests of the scholar and archaeologist with that of the designer. More and more the mantle of the architect passed to those men who could best unearth, sketch, and imitate the remains of classical Roman buildings. This new concept of the function of the architect had its effect on the status of architecture in the professional world by attracting into its ranks men from a higher social class than had been the case in the Middle Ages. In a felicitous expression Professor Ackerman has noted that 'if a man was not a gentleman before practising architecture, he became one after'.[2] Furthermore, the architect was no longer expected to serve an apprenticeship in the

[1] See John H. Harvey, *Henry Yevele, c. 1320 to 1400: The Life of an English Architect*, 2nd ed. (London, 1946), 28–29.

[2] James S. Ackerman, 'Architectural Practice in the Italian Renaissance', *Journal of the Society of Architectural Historians*, xiii, no. 3 (1954), 3.

mason's or carpenter's craft. For instance, Antonio da Sangallo the Younger was the only important architect of Renaissance Rome who rose from the building trades, and he was able to do so only through strong family connexions and the fact that he had worked under Bramante at the Vatican.[1] Instead of serving an apprenticeship in the building crafts, the architect was more often originally a painter or sculptor who turned to architecture only later in life. As a result, Italian Renaissance architects often did not have a thorough knowledge of the structural problems of building; they were required to leave these problems to the building craftsmen who had the necessary experience for solving them.

This professionalization of architecture, with its separation of the function of design from that of execution, was paralleled by another development which occurred during the Renaissance, namely, the separation of the functions of the architect from those of the military engineer. In the Middle Ages these roles were normally combined in the same occupation; medieval master masons and master carpenters were expected to build castles as well as cathedrals, so that there was little if any distinction between the military engineer and the architect. But during the sixteenth century in Italy there developed a separation between the architect concerned with the design of churches, palaces, and public buildings, and the military engineer who devoted himself entirely to the construction of fortifications. This change is reflected in the architectural literature of the period: whereas Alberti, Filarete, and Giorgio Martini had dealt with both civil and military architecture, a specialized literature on fortifications was produced in the works of Battista della Valle (1524), Giovanni Battista Zanchi (1554), Giacomo Lanteri (1559), Girolamo Cataneo (1564), and Girolamo Maggi (1564), to name but a few of the earlier writers.[2] The result was to create another profession—that of military engineering as apart from architecture.[3] Though of course there

[1] Ackerman, 'Architectural Practice', 3.

[2] For full bibliographical references to these early works on fortification, see Maurice J. D. Cockle, *A Bibliography of English Military Books up to 1642 and of Contemporary Foreign Works* (London, 1900), 197–216.

[3] The preceding was written before discovering the article, 'Military Architecture and the Radial City Plan in Sixteenth Century Italy', *Art Bulletin*, xlii (1960), 263–90, by Horst de la Croix, who had already reached exactly the same conclusions, based on a much wider reading of the Italian sources than I have attempted. De la Croix would even pinpoint this separation of military engineering from civil architecture to the period around 1550 (pp. 273–4). Cf. James K. Finch, *The Story of Engineering* (Garden City, N.Y., 1960), 118–19. Bertrand Gille, *Les Ingénieurs de la Renaissance* (Paris, 1964), while recognizing the development of

continued to be some overlapping of these professions in the work of particular individuals, by and large the military engineers came to be associated with a different background in their training and with a different outlook in their occupation. Sixteenth-century Italian architects were often painters or sculptors who approached architecture as another art form, and who looked to ancient Rome for the standards of their art. Military engineers, on the other hand, often had a background in military life; their occupation kept them tied more closely to the practical considerations of defence than to any aesthetic sense of their calling; and they faced the task of devising adequate fortifications against a destructive power which the Romans had never known—the cannon. Just as Italian architects of the fifteenth and sixteenth centuries developed a new style of architecture which soon spread over Europe, so Italian military engineers in the early part of the sixteenth century developed a new type of defence against the cannon, which likewise was adopted throughout Europe during the century.

From what little is known of the early career of John Rogers it appears that he was trained within the medieval tradition of the mason's craft. We find him first working as a stonesetter and carver at Hampton Court; within a few years he had become master mason at Guines. He was then appointed as surveyor of the works successively at Hull and Boulogne, and in this capacity he achieved eminence as one of the foremost military engineers in the service of Henry VIII. For several years after 1550 he served mostly as a technical consultant, but he ended his career as surveyor of works at Calais. Even this brief review of Rogers's career suggests that it was not entirely 'medieval' in character; if he began the normal career of a medieval master mason, he ended up rather differently, for he channelled his talents into the special role of the military engineer, in conformity with the specializing tendencies of the age. From the time of his appointment at Guines his services were largely devoted to the Crown in military engineering projects of one kind or another.

When this career is plotted against the changing conditions of the building professions in England and on the Continent, a rather interesting conclusion emerges. The professionalization of architecture in Italy in

specialization in the sixteenth century (p. 111), does not develop the point, for his study really culminates with Leonardo da Vinci, who died in 1519. W. B. Parsons, *Engineers and Engineering in the Renaissance* (Baltimore, 1939), failed to note this feature of Renaissance engineering (see especially p. 14).

turn produced a separation between architecture and engineering. In England, on the contrary, it would appear that by the middle of the sixteenth century engineering was emerging as a specialized profession before architecture had achieved the kind of professionalization already known in Italy. It has been said that Inigo Jones (1573–1652) was the first English architect in the full Renaissance meaning of that term.[1] If this is true, then it was three-quarters of a century after the new Italianate style of architecture was introduced into England before there emerged the Italianate type of architect in a man such as Jones.[2] On the other hand, by the middle of the sixteenth century men like Rogers and Lee were the full equivalents of the Italian engineers who were acquiring Europe-wide fame for their special skills and knowledge.

The more rapid development of engineering as a specialized profession appears to have been the result of Henry VIII's interest in providing England with a satisfactory network of military fortifications. By devoting so much of the money and energy of his Government to these large-scale projects, he made it possible for building technicians such as Rogers and Lee to specialize in fortification works alone. But there were other factors entering the picture, for it was not only the number and size of Henry's military projects that called for specialized building technicians, it was also the increasingly technical character of early modern warfare. The Italians were among the first to respond to this need by producing military engineers specially suited to the conditions of artillery warfare, and they were widely employed throughout Europe for the purpose of devising new fortifications able to resist the onslaught of the cannon. As we have seen, Henry himself employed some of these foreign engineers, but he also purposely cultivated the abilities of some of the English technicians in his own Office of Works. The result was that men like Rogers, Lee, and Cavendish emerged as qualified military engineers, who could readily take their place alongside their peers from Italy, Germany, and France.

The manner in which these men achieved their eminence as experts in fortifications is interesting, for it illustrates the English custom of adapting traditional practices and institutions to meet new conditions. Instead of

[1] Jenkins, op. cit. 42.

[2] On the problem of why this was the case, see John Summerson, *Architecture in Britain, 1530 to 1830*, 4th ed. (London, 1963), 24–25. Contrariwise, France had produced Philibert de l'Orme (*c.* 1505–77), the son of a French master mason who became one of the most famous French Renaissance architects of his day. The best recent study of de l'Orme is by Anthony Blunt, *Philibert de l'Orme* (London, 1958).

creating a brand-new position in the Office of Works which would be especially designed for military works, Henry and his ministers over the years merely adapted an older office to fill the need for a special kind of building technician. Since the fourteenth century the post of surveyor of works had been primarily an administrative one. As such, the surveyor was often responsible for supervising the payment of money, the acquisition of men and materials, and the checking of accounts to prevent misappropriation of funds and materials.[1] These more or less administrative and clerical duties were retained in the office during Henry's reign, as we have seen in the case of John Rogers, who carried a considerable administrative burden along with his other duties. But it was these other duties which the king imposed upon his surveyors of works that really transformed the character of the office, for the recipient of this office on fortification projects was expected to possess the technical knowledge and skills necessary for the construction of such works. Thus, several of the contemporary surveyors of works became the English equivalents of those foreign engineers—particularly Italian—who were so highly regarded as experts in fortifications.

As a specialist in fortifications the surveyor would have to be knowledgeable in the current theories and practices of military defence. Since he was responsible for supervising the construction of fortification works on a particular site, he would have to be conversant with the technical problems of construction as well as those of design. In his work he would be expected to display his skill as a draughtsman in the production of those plats that were becoming a vital part both of the building process and of communication between the builders and the royal patron. In the cases of the surveyors at Boulogne and Calais this skill involved them in diplomatic negotiations, where their ability to draw plats of the territories under negotiation made them valuable members of diplomatic commissions. The English surveyor thus faced a multitude of duties, for he was the one person immediately responsible for all of the technical and administrative elements directly connected with building works at a military post. To him were directed the technical problems that arose from below in the process of construction, and through him were channelled the administrative decisions and commands coming from above which might affect the character and progress of the works. Thus the surveyor stood

[1] H. M. Colvin, ed., *The History of the King's Works* (London, 1963), i. 168–9.

at the dividing line between the military and civilian personnel of the post. The hundreds of civilian construction workers were subject to his authority as the supervisor of their work, while he himself was responsible to the commander of the post because of the military significance of his work. In any show-down of authority the military commander was of course superior to the surveyor, as Rogers himself had to be reminded once or twice. Nevertheless, the surveyor had a very practical power at his disposal, for he possessed the technical knowledge for directing the works that very few military officials enjoyed. Consequently, as we have seen in Rogers's career at Boulogne, the military personnel could find themselves effectively frustrated if the surveyor either absented himself from the works or balked at their wishes and commands regarding the works. This frustration could be particularly galling if the monarch favoured his surveyor over his military commanders in matters dealing with fortifications, as Henry sometimes did at Boulogne.

Before his death in 1547 Henry had learned to put considerable trust in Rogers's judgement on various matters, particularly those dealing with military fortifications. He rewarded Rogers handsomely for this trust by paying his surveyor a high salary relative to that which other technical building officials received at the time. In addition he bestowed upon Rogers in 1544 the office of Clerk of the Ordnance, which was an unusual administrative post for a building technician in the service of the king. But then it reflects the close association of Rogers's career with the military life. Perhaps it also corresponds to the kind of reward granted to Richard Lee, Henry's other favourite English engineer, who became General Receiver of the Court of Wards and Liveries in the same year that Rogers was made Clerk of the Ordnance. Certainly Lee's reward was much greater in both position and income; it thus shows the relative status of the two men. Furthermore, Lee was knighted in 1544 for his part in a military expedition to Edinburgh and for the plan of the city which he showed to the king upon his return.[1] He holds the distinction of being the first English engineer to be knighted, and he represents thereby the increasing importance of the military engineer and the improved social status of the designer in sixteenth-century society. John Rogers never obtained the social eminence of Sir Richard Lee, but if Lee deserved to be knighted for making a plat of Edinburgh, Rogers surely qualified on similar grounds

[1] John Harvey, *English Mediaeval Architects* (London, 1954), 162; *D.N.B.* xii. 810.

for the numerous plats which he produced for the perusal of his king. Perhaps if Henry had lived longer Rogers might have received a like reward for his services at Boulogne.

In evaluating Rogers's career, we see that he stood somewhere between the positions of the medieval master builders and the modern professions of the architect and engineer. This can be seen both in his economic rewards and his status amongst government officials, and in the position which he held as a surveyor of works. But it can also be seen in the technical knowledge and skills which he displayed in fulfilling his duties as a surveyor. These show, on the one hand, the proficiencies and deficiencies of his heritage from the craftsmen of the Middle Ages. On the other hand, his work reveals certain newly achieved technical advances that were to play an important part in the making of the modern architect and engineer. By investigating certain aspects of sixteenth-century technics which are reflected in Rogers's work we can make the meaning of these statements clearer.

The date at which Rogers first became prominent in royal building works, around 1540, was a significant one in the history of English fortifications. Henry VIII had recently begun a campaign to fortify the southern coast of England as well as the English territories on the other side of the Channel. But there was not only a spate of military building activity at this time; the period also witnessed important advances in the art of fortification. In both developments Rogers played a significant role. As master mason at Guines and then surveyor of works at Hull and Boulogne he was responsible for the technical supervision of some of the more important military building projects of Henry VIII. Furthermore, study of the plans of these works suggests that Rogers was in the forefront in the development of the new methods of fortification which made their appearance in England at this time.

The new style in fortification was developed in Italy during the early part of the century.[1] Its rapid diffusion into other countries was the result, first, of the inherent merits of the new system, but it was also due to the ubiquity of Italian engineers on military projects throughout Europe

[1] This is not to say necessarily that the new style was *invented* in Italy; the best introduction to this vexed question is now J. R. Hale, 'The Early Development of the Bastion: An Italian Chronology *c.* 1450–*c.* 1534', *Europe in the Late Middle Ages*, ed. J. R. Hale, J. R. L. Highfield, and B. Smalley (London, 1965), 466–94, where the extensive literature on the subject is systematically reviewed.

during this period.[1] In addition, information was available to military engineers through the publication of numerous books which described and explained the new system of fortification.[2] The changes in design had been necessitated by the increasing efficiency of artillery, which had made obsolete the medieval castle with its high curtain walls flanked by even higher towers. The purpose of these high walls was to make it difficult to hurl missiles over them or to scale them with ladders. But these advantages had been lost with the greater power of cannons, which could fire cannon balls on a low trajectory with tremendous force and thus bring the tall curtains and towers down upon the heads of the defenders. The new style therefore called for lower and more solid structures which would provide less conspicuous targets to enemy gunners, and which, even when hit, would not come tumbling down. Thus the new fortresses became squat and almost solid structures, composed largely of earthen embankments encased within a battered masonry revetment.

These changes in general elevation and wall construction were made so that the fortress could better absorb the fire-power of the enemy artillery. In addition, the new style called for changes in the ground plan which would give certain advantages to the gunners in the fortress when firing on the advancing enemy infantry. Series of ditches and escarpments were placed around the fortress to impede the advance of the infantry and to provide a clear shot into their ranks as they moved towards the fortress. The fortress itself was designed so that gunfire could be directed at all angles in order to provide complete coverage over advancing enemy forces, no matter from what direction they might be coming or how close they might be to the fortress. This was achieved through a combination of angular bastions—placed all around the fortress—which projected away from the curtain walls to give the fortress a kind of star-like pattern.[3] While the curtain itself provided for straight-away fire-power, the design of the adjacent bastions allowed the gunners to maintain a crossfire over the field. Furthermore, the placement of these bastions was adjusted so

[1] The extent to which Italian engineers found service in foreign countries is shown in the three massive tomes of Leone A. Maggiorotti, *Architetti e architettura militari*, in *L'Opera del genio italiano all'estero*, ser. 4: *Gli architetti militari* (Rome, 1933–9), which in spite of their volume, yet do not include England and France.

[2] Cf. Cockle, op. cit. 197–216. Maggiorotti, op. cit. 22–45, and Horst de la Croix, 'The Literature on Fortification in Renaissance Italy', *Technology and Culture*, iv (1963), 30–44, provide running commentaries on these early technical treatises by Italian and other authors.

[3] See, for example, the plan of the French fortress across the river from Boulogne, Pls. 20 and 21.

that the face of each was protected by guns firing parallel to the face from points on the adjacent curtains or nearby bastions. Finally, one of the most significant features in the new style lay in the design of the bastion's flanks, which were slightly recessed into the bastion itself. This recess was produced by projecting from the shoulder of the bastion an orillion or 'ear' (Italian, *orecchione*).[1] The orillion protected the gunners in the recessed flank while allowing them to fire very close to the bastions and curtains. Thus even if the enemy artillery effected a breach somewhere in the defences, the enemy infantry would be prevented from entering it because of the murderous crossfire from the flankers of the adjacent bastions.[2]

By 1541 some of the principles of this new defensive system were being applied to the new fortifications at Guines in the Calais Pale and at Hull in Yorkshire. Although these works have since been destroyed, several contemporary drawings show to what extent the new system was utilized by the engineers who designed these projects. One of the plats, an elevation of the south side of Guines Castle (Pl. 4), depicts the typically medieval round towers and intervening curtains before the reconstruction of the castle had begun. Another drawing is a plan of the castle which reveals the designer's attempt to adapt the castle to the conditions of contemporary artillery warfare (Pl. 3). The angular bastions were there, but they were modified by the addition of semicircular projections which created awkward angles for the gunports. If the designer was attempting here to introduce the new system of fortification, he clearly misunderstood the basic principles of that system.

Another design for new bastions for the castle and town walls (Pl. 2) represents a considerable advance over the previous plan. Here the awkward combinations of triangles and semicircles have been replaced with symmetrical pointed trefoils. These bastions were still not designed strictly according to the Italian scheme, but they were intended to serve the same functions. Note the design of the flanks, which gave a protected placement for guns that would have a clear shot down the length of the curtains and past the face of adjacent bastions. Furthermore, the placement of the gunports in the trefoils made possible a fairly adequate crossfire over the field. However, the trefoil plan reveals a basic weakness that

[1] Cf. the fort on Plat C of Ambleteuse, Pl. 17.

[2] For more elaborate discussions of the design and function of the new bastioned fortification system, see de la Croix, 'Military Architecture', 265–7, 280–4; and Hale, op. cit. 474–7.

the new type of bastion was intended to overcome: the angles between the lobes of the trefoils contained a small but crucial area that could not be raked with crossfire from the curtains or adjacent bastions. A later drawing of the castle (Pl. 5) demonstrates yet another departure from the new system, for here it is seen that the bastions were multi-storeyed, having two levels of gunports within the bastions, plus gun embrasures in the parapets on top of the bastions. In short, there was no attempt here to build the low-lying, solid bastion—composed largely of earth with masonry revetments—that was one of the hallmarks of the new defensive system against the destructiveness of the cannon. Instead, these bastions at Guines appear to have been masonry structures with internal levels from which defenders could fire upon besiegers in a manner that was more medieval than modern.

Much the same kind of fortification was begun at Hull during the winter of 1541/2. A contemporary plat (Pl. 8) shows that the bastions at either end of the new fortifications were based on the same trefoil plan as the new bastions at Guines, while a later print by Hollar (Pl. 7) indicates that the elevations of the Hull bastions were also very similar to those at Guines.[1] But there were some differences in the designs on these contemporary plats (Pls. 2 and 8): the bastions at Hull were enclosed on all sides and had interior partitions, whereas at Guines the gorges were open and there were no interior partitions. Also, the axes of the Hull bastions were on the same line as the curtains, while those at Guines were at right or obtuse angles to the curtains. Even with these differences, however, it is clear that the works at Hull were based directly upon the designs which had been developed at Guines.

There can be little doubt that the designs at Guines were the product of the two English technicians responsible for the new works there, namely, the surveyor of works, Richard Lee, and the master mason, John Rogers. Documentary evidence shows that Lee was intimately involved in planning the works, while there is good reason for thinking that it was Rogers who drafted the plats that have provided the most vital information on the character of those works. But where did these two men get the ideas on fortification that are expressed in these plats? Since they were English

[1] It is curious that Hollar portrays the lobes of the trefoils as semicircular, when they were in fact pointed, as shown by a less detailed, but in this respect more accurate, view by John Speed, *The Theatre of Great Britaine* (London, 1611), following p. 81.

engineers working for an English king who was at that moment deeply involved in fortification projects elsewhere than at Guines, it is natural to look first at those other projects as possible sources for the design of the new bastions at Guines.

Henry VIII's string of forts along the southern coast of England, begun in 1538, is justly famous as 'the most extensive scheme of work of its kind undertaken in England until the last or even the present century'.[1] This scheme is interesting not only because of its size, but also because it reveals a concerted attempt to devise a type of fortification that would more adequately meet the demands of modern artillery warfare. The result was the construction of several forts which show some elements derived from continental ideas on fortification, but which in their overall effect differ considerably from contemporary developments in Europe. It is not necessary here to make a detailed analysis of the origin and development of all the features on these new forts.[2] We may be content merely with a description of those elements which show similarities to the new works at Guines and Hull.

Of the several forts begun during and soon after 1538, those which reveal the closest parallels in elevation to the new work at Guines and Hull were Walmer and Sandown, two of 'the three castles which keep the Downs' off the coast of Kent between Dover and Sandwich. This remark is motivated by a perspective view of Sandown, published in 1736 by Samuel Buck, which shows striking similarities between the bastions of Sandown and the new works at Guines and Hull as seen in Plates 5 and 7.[3] Unfortunately, Sandown Castle has entirely disappeared since this eighteenth-century engraving was made. We might take consolation in

[1] B. H. St. J. O'Neil, *Castles and Cannon: A Study of Early Artillery Fortifications in England* (Oxford, 1960), 49.

[2] As O'Neil has pointed out (ibid. 63–64), a quest for Continental precedents would lead to Germany and, specifically, to Albrecht Dürer, whose *Etliche underricht zu befestigung der Stett Schloss und Flecken* (Nürenberg, 1527) contains descriptions and illustrations of fortifications that preview certain elements in Henry's forts (see especially signatures Ci^{v}, Ciii^{v}, D, and F). Dürer's book might have been studied independently by the English technicians working for Henry, although a more likely source of contact with Dürer's (?) ideas would have been through Stefan von Haschenperg, who was employed as 'devisor' of Sandgate Castle in 1539. Cf. B. H. St. J. O'Neil, 'Stefan von Haschenperg, an Engineer to King Henry VIII, and his work', *Archaeologia*, xci (1945), 140–1.

[3] S. and N. Buck, [*A Collection of Engravings of Castles, Abbeys, and Towns in England and Wales*] (London, 1736), ii. eleventh collection, no. 20. The engraving is reproduced in A. D. Saunders, *Deal & Walmer Castles*, 'Ministry of Public Building and Works Guide Book' (London, 1963), 18.

the facts that Sandown's sister castle, Walmer, was practically an identical twin, and that it has survived. But Walmer has undergone drastic modifications during four centuries of occupation by military and government personnel, so that its present appearance is quite different from what it was originally. In order, therefore, to make valid comparisons with Guines and Hull, we must attempt to reconstruct the original plans and elevations of Sandown and Walmer. For this purpose we have the present structure of Walmer and the perspective views of S. and N. Buck, as well as some eighteenth-century Board of Ordnance plans and sections of the two castles.[1]

Both Walmer and Sandown were constructed on a concentric plan, with a round tower encircled by a ring of four semicircular bastions and short connecting curtains. Separating the tower from the outer ring was an open court or passage, while beyond the ring a moat surrounded the entire castle. Gunfire could be directed all round the symmetrical castle from four tiers of guns—one atop the central tower, and three in the outer bastions. It is the arrangement of the gunports in the bastions which catches our attention, because the new bastions at Guines and Hull also contained three levels for guns (cf. Pls. 5 and 7). The lowest level of gunports at Walmer and Sandown was placed within the moat. Thanks to the large number of these ports—there were seven in each of the four bastions and one in each curtain between the bastions—the gunners at this level were able to cover all points within the moat. These thirty-two gunports were connected by a continuous gallery inside the enceinte of bastions and curtains.[2] Access to this gallery was by means of steps leading down from the inner court of the castle. The second tier of gunports was on the level of this court. Three of the bastions each contained three large ports symmetrically arranged to provide the widest coverage from the semicircular bastions. The fourth bastion contained only two ports, for the third opening was used as the entrance to the castle from a drawbridge which extended across the moat. The guns at this level fired across the top of the moat, and by the relative size of their gunports, they must have been the largest in the castle. The bastions were originally hollow, although the one with the entrance gate may have been roofed and used for

[1] These latter have been reproduced in W. L. Rutton, 'Henry VIII's Castles at Sandown, Deal, Walmer, Sandgate, and Camber', *Archaeologia Cantiana*, xxiii (1898), 26–27; and O'Neil, *Castles and Cannon*, Pls. 13 and 16.

[2] See the basement plan of Walmer Castle in Saunders, op. cit. 32.

a gatehouse. Furthermore, the gorges of these hollow bastions were left unwalled, presumably so that anyone who captured a bastion could not defend himself within it against those who still held the central tower overlooking the outer bastions and curtains. A third tier of guns could be fired from the top of the thick walls of the bastions. Here the guns and gunners were protected by a curved parapet that contained embrasures through which the guns could be fired. There appear to have been seven embrasures in each of the bastions, corresponding thereby to the seven gunports within the level of the moat.[1]

By the time the Board of Ordnance plans were drawn in the eighteenth century, at both Walmer and Sandown the two bastions on the seaward sides of the castles had been filled solid, thus eliminating the use of the three gunports on the second, or inner court, level. A stone platform was laid on the fill, and from there guns could be fired through the seven original embrasures in the parapet.[2] Later these embrasures were removed and the low narrow parapets supplied which still exist at Walmer. As a consequence of these changes, the original features at Walmer are hardly discernible, but the hollow bastions, the thick walls, and the embrasured parapets are still to be seen at St. Mawes Castle, Cornwall, one of Henry's forts begun in 1540. However, at St. Mawes the bastions contain only two tiers of guns. Since the castle was located on a hillside it had no moat, and thus it required no tier of guns within the level of the moat. As a result it presented a rather different aspect from both Walmer and Sandown, which were placed inside moats in low-lying areas.

Pictorial evidence shows the new bastions at Guines and Hull with a lower tier of gunports similar to those at Walmer and Sandown (see Pls. 5 and 7). But the moats around the bastions at Guines and Hull were water-filled, so that this level of gunports was somewhat higher than at the two sister-forts in Kent. There is another interesting point to note on the plan of the bastions at Guines (Pl. 2), where internal galleries connecting these gunports are shown, similar to those still to be found at Walmer. It will be noted further that this plat shows the bastions as being hollow, with no walls across the gorges, like the bastions originally were at Walmer and Sandown. On this point, however, there is conflicting evidence. The

[1] See especially Buck's view of Sandown.

[2] One of the Board of Ordnance plans of the ground floor at Walmer shows these embrasures, although they were actually in the next floor level, and this apparently misled O'Neil (*Castles and Cannon*, 59; cf. Pl. 16*b*) into thinking that these two bastions on the seaward side contained seven gunports at ground level.

later view of Guines Castle (Pl. 5) represents the new bastions with a roof just below the level of the embrasured parapets; though the drawing is ambiguous in this respect, there may have been internal floors as well as walls across the gorges of the bastions. It is possible that the decision was made to include these features *after* the plan of the town walls (Pl. 2) was drafted. At any rate, when Rogers went to Hull he certainly added these features to the two end bastions, for they are indicated on both the contemporary plats and the later view by Hollar (Pls. 7 and 8). In these particular respects, then, Lee and Rogers moved away from the designs of the bastions at Walmer and Sandown. There were two other major changes. One was that the Guines bastions were pointed trefoils in plan, whereas those of Walmer and Sandown were semicircles. The other was that the Guines bastions had flankers connecting them to the curtain, while those at Walmer and Sandown did not. Interestingly enough, parallels to the first of these two elements can be found in other contemporary English forts built under royal patronage. Sandgate Castle in Kent, Hurst Castle in Hampshire, and St. Mawes Castle in Cornwall were built on plans that were approximately trefoil in shape, and that may have inspired the more precise trefoils of the new bastions at Guines. Furthermore, the bastions at Hurst were enclosed by means of walls across the gorges, floors and partitions within, and a roof overhead.[1]

Most of the features of the Guines bastions could thus be found in one or more of the other forts under construction on the southern coast of England after 1538. But one element of these bastions was distinctive, and that was the use of flankers. (See Pls. 2 and 5.) O'Neil recognized the distinctive character of the bastions at Guines and Hull and suggested that they may have represented a development towards the new type of bastioned fort that was independent of, or at least without direct dependence upon, continental models.[2] This is an interesting possibility, and the foregoing analysis allows a somewhat more precise formulation of the idea. The analysis has shown the extent to which Lee and Rogers utilized elements drawn from other contemporary English forts; on the other hand, there is little precedent in these English works for the special character of the flankers at Guines and Hull. This latter element must then either have been an independent invention by Lee and Rogers, or

[1] See the Board of Ordnance plans and sections of the castle in O'Neil, *Castles and Cannon*, Pl. 17*a*.

[2] Ibid. 68.

else they got the idea from continental examples or continental engineers who were familiar with the new Italian system of fortification.[1] There is little possibility of categorically demonstrating which alternative was the case, but the greater likelihood is that Lee and Rogers had learned something of the principles of the new system through the foreign, and especially Italian, artists and technicians whom Henry VIII employed in his service in one way or another.[2] The new bastions at Guines represent the attempt of the English engineers to formulate these principles within the framework of the designs currently being developed on other royal English fortification projects. That their own thinking was at an experimental stage is suggested by the differences in the plans for the new bastions at Guines as shown on Plates 2 and 3. This hypothesis accounts for the fact that the bastions at Guines contained flankers that were rather advanced in design, while other fundamental elements of the new system failed to be utilized. Thus the Guines bastions were not the low, solid, earth and masonry structures that the new style called for, but were instead multi-storeyed and built largely of masonry.

The military works designed by Lee and Rogers at Guines and Hull during 1541 involved a combination of features drawn from the Italian-type bastioned fortification and from the concentric forts which Henry VIII was currently building on the southern coast of England. When Rogers became surveyor of works at Boulogne, however, these English precedents were set aside, and the new fortifications were constructed strictly according to the Italian scheme. Survey I of Boulogne (Pl. 19), an early plat *not* drawn by Rogers, shows 'The Young Man' as a rectangular fort with 'medieval' round towers at the corners, while the bastions of the pentagonal French Fort d'Outreau are spade-shaped rather than triangular. But all of the plats of Boulogne and Ambleteuse that were drawn by Rogers (Pls. 15–17, 20–21) reveal the use of a fully articulated system of projecting angular bastions, with the flanks protected by orillions and with the bastions and intervening curtains surrounded by ditches or moats—all of which were patterned after the Italian style. In short, a major change

[1] They could hardly have come upon the idea in printed books discussing the new system, for these did not begin to appear until after 1550: Horst de la Croix, 'Literature on Fortification in Renaissance Italy', 39.

[2] On these men, see R. W. Carden, 'The Italian Artists in England during the Sixteenth Century', *Proceedings of the Society of Antiquaries of London*, 2nd ser., xxiv (1911–12), 171–204; and A. E. Popham, 'Hans Holbein's Italian Contemporaries in England', *Burlington Magazine*, lxxxiv (1944), 12–17.

in English military fortifications had taken place by the time the English works at Boulogne were fully under way. That change can be documented in the plats relating to the works at Guines, Hull, Boulogne, and Ambleteuse, where the fortification systems can be seen drawing closer, step by step, to the new style which engineers in Italy had developed only a few years earlier. Furthermore, the results of this present study have shown that John Rogers was not only directly involved in the supervision of these particular works; he was also the draughtsman of the plats which constitute the best surviving evidence on the nature of these works. Thus he appears to have been a central figure in the transformation of English fortifications during the period in question. This is not to say, of course, that he was entirely responsible for the adoption of the new style. But certainly he was one of the leading English engineers, along with Sir Richard Lee, who realized the merits of the new system, and who adopted it and adapted it to the needs of English warfare. It is only rendering justice to him and Lee to recognize that by 1546 they were quite abreast of the latest continental developments in the science of fortifications.

VIII

JOHN ROGERS AS A PLATMAKER

ROGERS was held in high esteem by his contemporaries as a platmaker, and those plats of his which are extant give some notion of the work upon which his reputation was based. It is therefore possible to make a detailed analysis of his skill as a draughtsman and to evaluate his work within the context of contemporary techniques in architectural and topographical drawing. The first point to note in this respect is that Rogers's drawings cannot always be strictly classified as either architectural or topographical, for both of these categories apply to some of his plats. This characteristic is understandable when one remembers the reasons for which most of the drawings were produced, namely, to inform the king and his ministers of the progress of English and enemy fortification works, or else to instruct them in the lie of the land around those projects. Since a single plat could do both of these things if executed on a sufficiently large scale, there was no need to distinguish between architectural and topographical drawings. This is the case with all of the extant drawings produced by Rogers at Boulogne, and with several of those drafted by him at Hull. On the other hand, Rogers did sometimes execute drawings that were more strictly architectural in character, such as the plats of Guines Castle and Hull Manor. We may therefore use these as the basis for discussing his architectural draughtsmanship; later we shall concentrate on the other drawings for an analysis of his work as a topographer.

Among Rogers's extant plats there are represented three different types of architectural drawings: the plans of the castle and town walls at Guines and the plans of Hull Manor (Pls. 1–3, 10–12); the elevation of Guines Castle (Pl. 4); and the bird's-eye view of Hull Manor (Pl. 13). It will be remembered from the description of the historical context in which these drawings were produced that most of them were not, strictly speaking, working drawings intended for use in building construction. Instead they were intended primarily for the eyes of the king and his ministers, who could discern from the drawings the present state of the building or the proposed changes to be made in it.

In providing plats for the perusal of the king rather than for use in construction, Rogers was working within the medieval tradition of the mason's craft, for it appears that most architectural drawings surviving from the Middle Ages were not technical working drawings, but were plats devised for the benefit of the patron, who wished to see at an early stage what his building would look like upon completion. Even those medieval drawings which were intended for actual use in construction were not working drawings in the modern sense of the term, for they lacked two essential elements of the modern blueprint, namely, dimensions and details (the latter showing, for instance, the jointings of the stones, or cross-sections of walls, window jambs, etc.). This point must be kept in mind when considering certain late medieval contracts where it was stipulated that the building would be constructed 'according to the plat'.[1] This could have meant no more than that the building, upon completion, would closely resemble the drawing (whether a plan or elevation or both) which accompanied the contract. It cannot have meant that the building would be built exactly as detailed in the drawings, because those details simply were not there. The details had either to be supplied directly by the technical supervisor of construction or else left to the discretion of those craftsmen who possessed enough knowledge and skill to be able to work out such details on their own.

Although Rogers's architectural plats lack the detail and precision of modern blueprints, still his drawings reveal certain advances over the efforts of earlier draughtsmen. First, his use of dimensions—such as those given on Plats A and B of Hull Manor—was an important step in the development of technically satisfactory instruments for building construction. Without these stated dimensions the builders were required to seek this information directly from the architect, or else they had to rely upon some previously accepted system of proportions in order to arrive at the specific dimensions of a wall, a room, or a building. Secondly, the fact that all three plans of Hull Manor were drawn on the same scale indicates an increasing awareness of the value of maintaining a given scale when making more than one drawing of a given structure. It is clear that

[1] For references to such instances, see L. R. Shelby, 'The Role of the Master Mason in Mediaeval English Building', *Speculum*, xxxix (1964), 390.

Rogers recognized the importance of providing a scale for his drawings, for he either inserted dimensions or stated the scale on practically every plat that he drew. Finally, the provision of an elevation and/or a bird's-eye view in company with the floor-plans was a move towards the creation of a 'set of blueprints' which could provide different points of view of the building, so that the builder could solve technical problems without constant recourse to the architect.

It is not to be supposed that Rogers was the originator of these advances in technical drawing. His techniques do no more than reflect the current practices in architectural drawing in England around the middle of the sixteenth century. But this raises an interesting and very complex question: whence had come these techniques of architectural drafting that were practised in Rogers's day? In answer to this question it may be noted that there are extant several beautiful architectural drawings produced by English master masons in the late fifteenth and early sixteenth centuries.[1] These indicate that architectural drafting had become a recognized part of the late-medieval master mason's stock-in-trade, and that the acquisition of that skill was a necessary part of his education. Thus there is no need to assume that Rogers's skill in architectural drafting could have been acquired only in some special training apart from that which he would have received in becoming a master mason.

This is an important point and it should be emphasized. The late-medieval tradition of the mason's craft, of which Rogers was a direct heir, was itself producing advances in the techniques of architectural drawing, and these advances are seen in the work of men like Rogers. If that medieval tradition had been allowed to continue uninterrupted, presumably these advances would have continued to be made. Furthermore, it is clear that architectural drawings were becoming more and more important in the processes of building construction during the late Middle Ages, and this tendency too would have continued to develop into modern times within the traditions of the building crafts. But the medieval tradition was

[1] For examples one may cite the well-known elevations of King's College Campanile (B.M., Cott. MSS., Aug. I. i. 3), of King's College Chapel (Aug. I. i. 2), and of the sepulchral monument of Henry VI (Aug. II. i), reproduced in John Harvey, 'Early Tudor Draughtsmen', *The Connoisseur Coronation Book* (London, 1953), nos. II–IV; as well as the lesser-known medieval drawings in the Smithson Collection of the R.I.B.A. Library (Box 3 AE 7/15, 16, and 17), reproduced in Mark Girouard, 'Three Gothic Drawings in the Smithson Collection', *Journal of the Royal Institute of British Architects*, 3rd ser., lxiv, no. 1 (1956), 35.

interrupted in northern Europe during the sixteenth century by the introduction from Italy of classical forms of architecture, which in time swept away the architectural forms of the Middle Ages. This interruption had its impact not only upon architectural style, but also upon techniques of architectural drafting, as well as upon the use of architectural drawings in building construction.

With the revival of classical architectural forms in Italy, the antiquarian/archaeological interests of the scholar/architects who were responsible for that revival called forth increasingly detailed and exact copies of classical models in the form of architectural drawings. In the process, the measured drawing came into being, as well as more varied kinds of drawings—plans, elevations, sections, and perspective views of those ancient monuments which had survived. With the increased proficiency of architectural draughtsmen these measured drawings not only provided copies of the ancient models for the edification and delight of contemporary connoisseurs of classical architecture; they also opened up the possibility of handing a drawing to a working mason with the expectation that few if any further directions would be necessary for the exact translation of that two-dimensional drawing into a three-dimensional structure in stone. Be it noted, however, that this *possibility* was not fully realized all at once, even in the hey-day of the Italian Renaissance of the sixteenth century. Professor Ackerman, after a study of Renaissance architectural drawing, expressed his surprise over the fact that even the Italian architects of the early sixteenth century did not draft fully articulated sets of building drawings, but instead relied mostly upon measured plans and details which were to be supplemented by word-of-mouth instructions to the working masons.[1]

In short, the techniques of Renaissance architectural drafting as well as the use of drawings in construction did not constitute a categorical break with the medieval past. But the Renaissance did have the effect of accelerating advances in technique and of increasing the use of drawings in actual construction work. As the medieval traditions of design withered away before the impact of classical forms, the working masons found themselves ignorant of the principles underlying those forms, so that they became more dependent upon the architect to supply them with the

[1] James S. Ackerman, 'Architectural Practice in the Italian Renaissance', *Journal of the Society of Architectural Historians*, xiii, no. 3 (1954), 8.

details for their work. These details could best be presented in architectural drawings, which increasingly served as the intermediaries between the working masons and the scholar/architect or gentleman/architect who had familiarized himself with the principles of classical architecture. There was thus an increased incentive for perfecting the techniques of drawing, in order to ensure the correspondence between the work of the masons and the intentions of the architect. While those techniques were being perfected, however, the medieval traditions of the building crafts supplied whatever technical knowledge and skills were necessary to overcome the deficiencies remaining in the architectural drawings. As those deficiencies were gradually ironed out, the craftsmen became more and more dependent on the architect to supply the detailed answers for technical problems through the increasingly effective architectural drawings. This reciprocal development ultimately produced the modern situation, in which the construction worker's role in the design of a building is reduced to zero, and his only problem is how to put the building materials in their proper place as shown in the blueprints.[1]

Even if the Renaissance did not provide a radical break in the development of architectural drawing, still the sixteenth century witnessed an important turning-point in the story, thanks to the advances made by the scholar/architects who were interested in developing techniques for making correct copies of classical models. Where does John Rogers stand in relation to these advances of the sixteenth century? Among his English contemporaries his work compares well. His floor-plans reveal a precision of line and firmness of hand that account for his reputation among English technicians as a platmaker. Plan C of Hull Manor (Pl. 12) best displays the careful draughtsmanship of which he was capable. This drawing alone puts him in the front rank of English architectural draughtsmen of his own day; it also compares favourably with the many extant floor-plans produced by the later and more famous Elizabethan and Jacobean architect and surveyor, John Thorpe.[2]

[1] This last statement is perhaps a bit exaggerated, for rarely are the blueprints of a building so detailed, or so correct and consistent in their details, that the workers have nothing to do but execute the plans. In cases where details are either missing or incorrect, the supervisor of construction may determine his own solution of the problem, unless it is of such proportions that the architect needs to be consulted.

[2] See *The Book of Architecture of John Thorpe in Sir John Soane's Museum*, edited with Biographical and Analytical Studies by John Summerson, Walpole Society, vol. 40 (Glasgow, 1966).

The elevation of Guines Castle (Pl. 4) is a workmanlike job, while the bird's-eye view of Hull Manor (Pl. 13) is a striking drawing, even if it is technically inaccurate in its projection. But it is precisely in this matter of projection that we find a clue for evaluating Rogers's skills. The drawing combines perspective with orthographic projection in a way that is similar to other late medieval architectural drawings.[1] The difficulties which Rogers had in representing converging rooflines, or in projecting the tower relative to the lower parts of the building, indicate that the tradition which had taught him skill in handling the drawing pen had not provided him with solutions to certain technical problems in architectural drafting. That is to say, Rogers was still hampered by some of the imperfect techniques of medieval drawing, for apparently he had not received the opportunity to profit from the technical advances made by Italian draughtsmen in the late fifteenth and early sixteenth centuries.

Comparison of Rogers's drawings with those of a contemporary French architect, Philibert de l'Orme, is instructive on this point. De l'Orme was himself the son of a French master mason, so that, like Rogers, he received his early training in the mason's craft. But he enjoyed the privilege of completing his training by studying in Italy, where he acquired a knowledge not only of the idioms of classical architecture, but also of contemporary Italian techniques of architectural drawing. Both of these accomplishments were illustrated in his famous *Premier Tome de l'architecture* (Paris, 1567) where he displayed his drafting skill not only in measured drawings of technical details, but also in plans, elevations, perspectives, and interior views of rooms, domes, archways, etc. The differences in the techniques of de l'Orme's and Rogers's architectural drawings are themselves the measure of the advances which had been and were currently being made by Italian draughtsmen and by northerners who had come under their influence.

From Rogers's architectural drawings we may turn to those plats which included both architectural and topographical features, and which may be used for evaluating his place in contemporary practices of topographical drafting. The sixteenth century witnessed important developments not only in architectural drawing but also in the art of making maps. The accumulation of geographical data in the 'Age of Exploration' was com-

[1] Cf. for instance the sepulchral monument of Henry VI and the Campanile of King's College, referred to in p. 147, n. 1 above.

bined with the increasing technical skills of cartographers to produce the beautiful regional and world maps of the latter half of the century—the most famous being those of the Flemish cartographer, Gerard Mercator. But John Rogers was only indirectly related to that story, for he was not a geographer or cartographer in the technical sense, but rather a surveyor-engineer who was commissioned to make topographical drawings of limited areas. In this connexion one may note the work of the contemporary English estate-surveyors, who performed the less exalted but none the less useful task of surveying and drafting plats of those estates which were passing into the hands of the gentry upon the dissolution of the monasteries. At first the efforts of these estate-surveyors were quite amateurish, but in time technical advances were made which laid the basis for the accomplishments of late sixteenth-century regional map-makers such as Christopher Saxton and John Norden.[1]

Rogers was himself part of a closely related tradition—that of providing detailed topographical maps of certain coastal havens, fortifications, and towns which were of particular military, political, or economic interest to the Crown. Henry VIII took special interest in such work, as indicated by the large number of drawings in the British Museum which date from his reign. For the task of mapping these various sites Henry turned to both foreign and English surveyors and engineers who were in the royal service. As surveyor of the works at Hull and then Boulogne Rogers was naturally drafted into the role of topographical platmaker, for this had become an accepted part of the surveyor's duties. But the numerous documentary references to Rogers's platmaking, as well as the large number of extant drawings which can be attributed to him, indicate that his skills in topographical drawing were held in particularly high esteem.

Even a cursory survey of the plats attributable to Rogers raises a problem which must be dealt with immediately, namely, that all of the drawings do not reflect exactly the same drafting style. On the basis of stylistic evidence alone, for instance, one would scarcely attribute the

[1] Edward Lynam, 'English Maps and Map-makers of the 16th Century', *Geographical Journal*, cxvi (1950), 11–12; this article, along with others pertinent to the present study, is reproduced in idem, *The Mapmaker's Art* (London, 1953). On the subject of sixteenth-century English cartography, see idem, *British Maps and Map-Makers* (London, 1944), 12–24; Sir H. G. Fordham, *Some Notable Surveyors & Map-Makers of the Sixteenth, Seventeenth, & Eighteenth Centuries* (Cambridge, 1929), 1–15; and R. V. Tooley, *Maps and Map-Makers*, 2nd ed. (London, 1952), 47–64 (the last with extensive bibliographies).

survey of Hull (Pl. 9) and Survey III of Boulogne (Pl. 21) to the same draughtsman. Yet, as we have seen, there is good documentary and handwriting evidence for attributing both drawings to Rogers. What is the explanation of their stylistic difference? The easiest answer is to suggest simply that Rogers changed his drawing style. But if one further recognizes that Survey III of Boulogne represents considerably more sophisticated techniques of topographical drawing than does the survey of Hull, then one may say that Rogers *improved* his techniques between the time that he drew the earlier and the later surveys. It is upon this assertion of Rogers's continuing improvement as a topographical draughtsman that the following analysis of his drawings will be based.

It might seem strange that Rogers's skill as a topographer underwent considerable improvement, whereas his techniques as an architectural draughtsman appear quite fully developed in his earliest extant drawings—those of Guines Castle. But in terms of Rogers's overall career this feature is not inexplicable. In his preparation for becoming a master mason he would have been taught the basic skills of architectural drafting as a necessary part of his building trade. But the opportunity (and necessity) of topographical drawing came to him only when he had achieved the position of surveyor of works at Hull, so that this skill was relatively undeveloped at first, but was gradually improved through his continuing production of topographical plats.

The plat of the Hull fortifications (Pl. 8) was primarily intended to show the plan of the new works across the river from Hull, and the topographical features of the drawing are crude indeed. The row of houses along the riverfront is a mere cartoon, and there are practically no details for the interior of the town. The Hull and Humber Rivers are distinguished from the surrounding land by a scroll-like pattern, which was only one of several devices used by Rogers to indicate bodies of water. The fact that these devices varied in the drawings which he produced over the years shows that he had been taught no special symbol for water and that he was free to experiment with various ways of representing it on his drawings.

A more truly topographical drawing than this plat of the fortifications is the survey of the area around Hull (Pl. 9). This drawing provides a much better indication of Rogers's skill as a topographer in the early days of his career as a surveyor of works. Rogers followed the not uncommon practice of that time in showing the rivers, roads, and terrain in plan, while the towns and villages are seen from a bird's-eye view. This characteristic of many sixteenth-century plats, while making them technically deficient because of the distortions created, nevertheless gives them a certain dramatic impact, not to speak of historical significance, because of the particular and sometimes quaint features which can be indicated thereby. Another feature of this drawing is that it is primarily drafted with pen and ink, although emphasis is provided by outlining the rivers, roads, and drainage ditches with a light wash. The Humber River, on the other hand, is indicated with strokes of a much darker wash. The use of wash distinguishes this drawing from all of the architectural plats by Rogers which we have previously considered, where no wash was utilized. Yet much the same effect of emphasizing the pen lines was created by Rogers's use of very fine hatching to outline the walls of the second-floor plan of Hull Manor (Pl. 12). It would appear that Rogers's use of wash was simply a faster way of producing this emphasis than the tedious method of hatching with pen and ink. It was a technique which he would continue to use, to some extent at least, in all of his topographical drawings.

Rogers's use of symbols was an important part of his development as a topographical draughtsman. One of the unusual features of this survey of the Hull area is the application of stippling to symbolize the dirt roads on 'the waye to Beverleye' and 'the waye to Cotingham'. Topographers of this period normally did not even indicate roads from one town or village to another, so that Rogers's advanced technique in the use of stippling is particularly interesting. The symbol for the low line of hills at the top of the drawing is a series of curving pen lines outlined with a light wash which shows the hills in silhouette. The topographical symbol for hills was one with which Rogers continued to experiment, and which therefore changed considerably in the development of his particular style. Furthermore, it will be noted that Rogers did not let the symbol for the hills speak entirely for itself, but added the words, 'The wolde', to indicate unequivocally that these were hills. This use of inscriptions to explain symbols that might be clear enough in themselves remained a common feature of Rogers's drawings. Indeed, this characteristic is so pronounced that it may be said to distinguish his work from that of other contemporary platmakers such as Richard Lee and Richard Cavendish.

Earlier analysis of the documentary evidence relating to Rogers's drawings of Boulogne and Ambleteuse suggested that Plats A, B, and C of Ambleteuse (Pls. 15–17) were drawn in late March and in early and late April 1546 respectively.[1] The 'Country of Guynes and Bolenois' (Pl. 14) may have been drafted by Rogers in late May, while Survey II of Boulogne (Pl. 20) was drawn in June and Survey III (Pl. 21) in August of that same year.[2] Stylistic analysis of these drawings will show a progression in technique which correlates closely with the temporal sequence in which the drawings were produced.

Plats A and B of Ambleteuse are rather hastily sketched drawings, which Rogers executed almost entirely in freehand. He relied primarily on pen and ink, although he used wash to outline in a very casual way the shores and cliff of the haven. Some of the details are sloppily executed, but the sureness of his hand when needed is vouchsafed by the unbroken lines which he drew to indicate the banks of the river and haven. Both drawings are bird's-eye views, from a position outside and above the harbour, looking eastward. Rogers's penchant for taking his point of view from above the sea and looking landward was continued in every one of his extant drawings of the Boulonnais. He need not have taken that position, of course; other draughtsmen assumed a different point of view, as illustrated in Survey IV of Boulogne (Pl. 22). As for the perspective of the two drawings, there is some improvement over that of the survey of the area around Hull (Pl. 9), although it is clear that Rogers had not entirely grasped the significance of showing both landscape and buildings in bird's-eye view, for on Plat B the fort is drawn in flat plan. A peculiar feature, the roofless houses, appears in these drawings and remains a characteristic feature of Rogers's style of platmaking. The appearance of this 'iconographical' oddity suggests that Rogers was being directly influenced by the drafting technique of other topographers, perhaps foreign to his own circle of friends and fellow engineers at Boulogne. It is possible that this particular feature can be traced to a specific drawing which also shows roofless houses—Survey IV of Boulogne. Earlier it was suggested that this was the plat produced by the French commissioners in the peace negotiations during the spring of 1546, which William Paget referred to as the fairest that he had ever seen. If this is the case, Rogers might well have seen the plat and got the idea of the roofless houses from

[1] See above, p. 79. [2] See above, pp. 99–102.

it. The meaning of the houses represented in this form is uncertain, unless they were intended to show the destruction due to war.

Plat C of Ambleteuse is a more carefully executed drawing of the fort and area around Ambleteuse. The topographical details are more precise, and presumably more correct, while the precision with which the pen lines are made can best be seen in the rigging of the ships lying in the harbour. This plat also shows advances in the careful use of wash to outline topographical features as well as the fort; it thus represents another step in Rogers's increasing application of this medium in his platmaking.

The 'Country of Guynes and Bolenois', on the other hand, is a fairly rudely drawn plat that was drafted with pen and ink, and with only a minimal use of wash. It is hardly more than a sketch, drawn in freehand, probably for the purpose of showing the towns, rivers, and forts in the Boulonnais and March of Calais. The roofless houses are prominent features of the drawing and provide a further clue that this plat was made by Rogers. There is an interesting attempt here to provide a single perspective from a bird's-eye view, although the attempt is not entirely successful. The hills, which are shown mostly in silhouette, do not conform exactly to this perspective; but they do come closer when the point of view is taken just outside the haven at Boulogne. When the drawing is studied from this angle, it will be seen that the very topography of the plat is drafted to focus attention on Boulogne. This construction reflects the importance of the town in the diplomatic negotiations for which the plat was drawn, as well as Rogers's own natural inclination to view the country round about from the vantage-point where he was engaged as surveyor of the works. Furthermore, it may be noted that the peculiar perspective is not unlike that of Survey IV of Boulogne, so that here again Rogers may have been imitating techniques which he encountered on this latter drawing.

Rogers's increasing skill in perspective rendering of topography is reflected in Surveys II and III of Boulogne. In both plats he assumes a similar position over the sea, looking landward. And in both plats the sand-hills are shown merely as silhouettes in the fashion made familiar by his plats of Ambleteuse. But the two prominent hills in the front and centre of the plats are depicted as 'sugar-loaf' hills, and there is an obvious attempt to render them with a three-dimensional quality. This three-dimensionality is partly achieved by Rogers's greater skill in using the drawing pen to indicate topographical features such as cliffs, seashore, valleys, and hills. In drawing these elements he combines precise and heavy lines with lighter and merely suggestive marks in order to give emphasis as well as to provide a sense of depth. In these respects Survey III shows advances even over Survey II. Particularly noteworthy is the fact that Boulogne and the English and French fortifications on Survey III are shown in bird's-eye view perspective, whereas these elements are merely drawn in plan on Survey II. But perhaps the greatest advance in Survey III is the skilful application of wash for giving roundness to the hills and depth to the valleys, as well as emphasis to the fortifications. This use of wash provides a fullness to the plat that is lacking on Survey II, which remains primarily a pen and ink drawing.

It is clear that by the time he produced Survey III of Boulogne Rogers had become as skilful in topographical as in architectural drawing. He showed distinct improvement in perspective rendering, where both topographical and architectural features were drawn from a single point of view high above the terrain. This improved perspective was aided by Rogers's increased ability to combine the technique of pen and ink drawing with wash, which enabled him to delineate technical matters of topography with precision of line, while at the same time obtaining a sense of depth and three-dimensionality—appropriate for a bird's-eye view—through shading with water-colour.

As a final consideration, it may be noted that a student of early English water-colouring has recognized the part played by Henry VIII's military draughtsmen in the development of this medium for topographical and landscape drawing. He not only finds the beginning of early English water-colouring in the sixteenth century; he further suggests that it was closely related to the work of the military engineers who produced plats of towns, fortifications, and havens for Henry VIII.[1] He bases his claim upon the large number of plats by these engineers that are preserved in the Cottonian collection of the British Museum. But since he suggests that most of these drawings were by foreigners, it is once again only rendering justice to point out that several of the drawings were in fact drafted by an English engineer who, in English circles at any rate, enjoyed quite a reputation as a platmaker. It is not our intention to overemphasize the

[1] Iola A. Williams, *Early English Watercolours and Some Cognate Drawings by Artists born not later than 1785* (London, 1952), 2.

role played by Rogers in the development of topographical drawing in this period of rapid improvement in the art, but we may at least note that here, as in the science of fortifications, there were some English engineers who were staying abreast of the latest developments on the Continent, and not the least important of these men was John Rogers.

APPENDIX A

The Handwriting of John Rogers and his Clerks

THE inscriptions on the drawings attributable to John Rogers pose an important problem in ascribing the drawings to him, for the handwriting of the inscriptions varies considerably. Because of this variety, it is beyond all likelihood that Rogers wrote the inscriptions on all these plats. But if the inscriptions were not by him, then this point might be used to argue that the plats themselves were not drafted by Rogers. In order, therefore, to secure the attribution of the plats to Rogers, an attempt must be made to explain how and why the inscriptions were written by hands other than that of the man who drew the plats. This admittedly will be a tricky business, because it will necessitate making distinctions between the handwriting samples of various sixteenth-century clerks. Since these professionally trained scribes could write almost identical hands with one another if they chose, or, on the contrary, could use different hands at different times, the opportunities for error in such an enterprise are great. Nevertheless, the importance of the handwriting factor in the attribution of the drawings to Rogers makes the effort not only worthwhile but necessary.

The first task is to establish the character of Rogers's own handwriting. There are numerous samples of handwriting which can be associated with him in one way or another; they may be listed in tabular form as follows:

Letters from John Rogers alone

(1) P.R.O., S.P. 68/13, no. 42. (Pl. 27)
(2) S.P. 68/14, no. 86. (Pl. 28)
(3) S.P. 68/14, no. 103.
(4) S.P. 68/15, no. 141. (Pl. 29)

Letters from John Rogers and others

(5) S.P. 1/166, f. 5.
(6) S.P. 1/166, ff. 92–93.
(7) S.P. 1/216, ff. 138–9.

Endorsements on drawings

(8) Plat C of Hull Manor: B.M., Cott. MSS., Aug. I. 84. (Pl. 25*a*)
(9) 'Country of Guynes and Bolenois'. Aug. I. ii. 75.

Accounts

(10) Clerk of Ordnance: B.M., Add. MS. 5753. (Pl. 26)

Inscriptions on drawings

(11) Plat A of Guines: Aug. I., Supp. 14. (Pl. 1)
(12) Plan of Hull fortifications: Aug. I., Supp. 4. (Pl. 8)
(13) Survey of Hull area: Aug. I., Supp. 20. (Pls. 9 and 25*b*)
(14) Plat A of Hull Manor: Aug. I., Supp. I. (Pls. 10 and 25*c*)
(15) 'Country of Guynes and Bolenois': Aug. I. ii. 75. (Pl. 14)
(16) Plat A of Ambleteuse: Aug. I. i. 59. (Pl. 15)
(17) Plat B of Ambleteuse: Aug. I. ii. 68. (Pl. 16)
(18) Plat C of Ambleteuse: Aug. I. ii. 73. (Pl. 17)
(19) Survey II of Boulogne: Aug. I. ii. 77. (Pl. 20)
(20) Survey III of Boulogne: Aug. I. ii. 53. (Pl. 21)

The letters provide the best clues to Rogers's own hand. All seven were signed by him, and there is such close agreement that one may feel certain these provide the authentic signature of John Rogers. Furthermore, a careful comparison of the signatures with the texts of the letters indicates quite clearly that letter no. 4 in the list above was written by Rogers himself. Compare for instance the majuscules *J* and *R* and minuscules *g* and *h* in the signature with the corresponding letters in the epistle. On the other hand, nos. 1 and 2 were written by hands that are clearly different from each other and from the autograph letter of Rogers. These letters were probably written by two of the clerks that Rogers was entitled to have as surveyor of the works at Boulogne.[1]

Having established with fair certainty that one entire letter was written by Rogers himself, we may use it as a basis for comparison with the other samples of handwriting which can be associated with him. The general appearance of the endorsement on the back of the Hull Manor drawing (Pl. 25*a*) is rather different from the autograph letter, and this impression is borne out by comparing again the majuscules *J* and *R* and the minuscules *g* and *h* in the two samples. An even greater contrast is seen, however, by comparing the entire word 'Rogers' of the endorsement with the 'Rogers' of the signature, where the formation and ligatures of all six letters are different. This at best ambiguous handwriting evidence is not relieved by the ambiguity of the endorsement: 'A new plat made by the same Rogers . . .'. If written by Rogers, why did he not say, 'A new plat made by me, John Rogers'? The cumulative effect of these difficulties suggests that the endorsement was not written by Rogers, but by some clerk who wished to put on record who was the draughtsman of the plat.

The endorsement on the 'Country of Guynes and Bolenois' (no. 9) is badly faded

[1] *L. & P.* xx. i. 39 contains a reference to Rogers's 'wardens and clerks'. For the record, it may be noted that letter no. 3 was scribed by the same clerk that wrote no. 1. I have been unable to trace the identity of the hand that penned the letters sent by John Rogers in conjunction with other officials (nos. 5–7).

and has been partly cut away, but it may be read as 'A plat of the contre of gy[nes and] bolinnoys'. One cannot speak with absolute certainty here because of the condition of the endorsement, but comparison of what remains with Rogers's holograph indicates that it was penned by Rogers himself. If this is the case, it constitutes another clue on the difficult question of who drew this plat.

Turning to Rogers's account as Clerk of the Ordnance that was submitted in July 1544 (no. 10), one can see with a glance at Plate 26 that at least one line of this page was written by Rogers, namely, the sentence 'Receyvyed by me John Rogers'. Further study reveals that three other hands are to be found on this page. At the top is the last part of the account, written in a formal Secretary Hand; below this is an illegible signature; and below that, the receipt of the account, written by a different hand and in a somewhat less formal Secretary than the account at the top of the page. It thus appears that two clerks were involved in preparing Rogers's account as Clerk of the Ordnance: 'Clerk A' drafted the major portion of the account, while 'Clerk B' wrote the receipt of account. The handwriting of Clerk B catches our attention because of its similarity to two handwriting samples already noticed—the endorsement on the Hull Manor drawing (no. 8) and two of the letters from Rogers while he was at Boulogne (nos. 1 and 3). Detailed comparison of Plates 25*a*–27 suggests that all of these documents were written by the same clerk, who would then have worked with Rogers at Hull, London, and Boulogne, over a period of time that began perhaps in 1542 and extended at least to April 1547, the date of letter no. 3.

It has become apparent that in his administrative duties Rogers relied heavily on his clerks to pen the various documents involved in his work, while he merely signed the completed draft. This practice has an important bearing on the problem of the inscriptions which are to be found on a number of drawings attributable to Rogers. The inscriptions on the plat of the Hull fortifications (no. 12) and the survey of the area around Hull (no. 13) show strong similarities. A number of letters could be cited, but the peculiarities of the majuscules *T*, *B*, and *H* in the inscription 'The lenght betwene the Bulwarke next the Humber . . .' on the one drawing, when compared with the same majuscules in the phrase 'The waye to Beverlaye' and the word 'Holdernesse' on the other drawing, leave no doubt that all of these inscriptions were made by the same hand (cf. Plates 8 and 25*b*). But by whose hand? Before attempting to answer this question, we should note a couple of points. In the first place, the inscriptions on the plans were written in a formal Secretary Hand—the predominant business hand at the time of these drawings.[1] Secondly, the close similarity of these inscriptions to some of the samples of

[1] On the general character of the Secretary Hand, see Hilary Jenkinson, *The Later Court Hands in England from the Fifteenth to the Seventeenth Century* (Cambridge, 1927), 57–62; Samuel A. Tannenbaum, *The Handwriting of the Renaissance* (New York, 1930), 21–22; and L. C. Hector, *The Handwriting of English Documents* (London, 1958), 57–58.

Secretary Hand illustrated in the first English copy-book[1] shows that they are the work of a well-trained scribe.

There is circumstantial and stylistic evidence that Rogers drafted these plats at Hull, but comparison of the inscriptions with Rogers's autograph letter reveals great dissimilarities in the handwriting. We are faced with three quite different alternatives. One is that Rogers did not execute the drawings, and the proof of this lies in the inconsistency of the handwriting evidence. But this view is based on the assumption that this inconsistency cannot otherwise be explained; furthermore, it fails to explain the positive evidence for Rogers's authorship of the plats. A second possibility is that Rogers did execute the drawings, but wrote the inscriptions on the drawings in a more formal Secretary Hand than he normally used for writing letters and his own signature. By the beginning of the sixteenth century it was not at all unusual for a person, as a part of his education, to learn to write both a formal and a cursive hand.[2] But of course it is difficult to know that Rogers had done so, since practically nothing is known of his early life and education. The argument for the second alternative suffers also from internal weaknesses. If Rogers was capable of writing the beautiful and formal Secretary Hand that is seen on the Hull drawings, it hardly seems likely that he would have used a cursive, informal hand in writing his autograph letter to the Privy Council. Furthermore, even if he could write two different scripts, one would expect to find at least some personal idiosyncrasies which would be common to both hands. Frankly, I have been unable to discern any such common characteristics.

The third alternative is that Rogers drew the plats, but allowed his clerk to write the explanatory inscriptions on them. While at first this might seem to be the least likely explanation, it can be supported by fairly strong arguments. For instance, there are similarities between the inscriptions on the Hull drawings and the handwriting of Clerk A, who scribed Rogers's account as Clerk of the Ordnance. In particular one should compare the formation of the majuscules on the drawings with those same letters in the proper names that are listed in the ordnance account.[3] The consistent formation throughout the documents of the minuscules *e* and *h*, which are considered to be test-letters for the Secretary Hand,[4] adds further weight to the case. Another point to consider is that the very formality of the hand argues for its having been written by a formally trained scribe, rather than by a technician like Rogers who would have been responsible for producing such a drawing, and who would not—in all likelihood—have acquired such a formal hand in his own education.

[1] John de Beauchesne and John Baildon, *A Booke Containing Divers Sortes of Hands* (London, 1571).

[2] S. H. Steinberg, 'Medieval Writing-Masters', *The Library*, 4th ser., xxii (1941), 1–2.

[3] The similarity of the rather unusual B in 'Bulwarke' and 'Beverlaye' (Pls. 8 and 9) to that in 'Thomas Bucklande' on fol. 121 of the account (not reproduced) is almost enough to clinch the argument by itself.

[4] Hector, op. cit. 58.

We conclude that, while Rogers himself drafted the plat, he caused the inscriptions to be written by one of his scribes—in this case the man whom we have called Clerk A. But this still leaves the delicate question of why Rogers should involve himself and his clerk in such a procedure. There are at least two possible reasons. One is that Rogers's busy schedule allowed him time to make the drawings, but inclined him to leave the inscriptions to be added by his clerk. The other possible reason—and the one which I favour—is that Rogers's own hand was nondescript, so that he preferred to have his plats inscribed by his professionally trained clerk, who could supply a formal hand that would measurably improve the appearance of those plats intended for the perusal of the king.

If it has been established that Clerk A did write the inscriptions on the Hull drawings, then we need not be disturbed when inscriptions on other drawings attributable to Rogers are also dissimilar to the handwriting of his holograph, for we may expect that he continued the practice of having his clerk write in whatever explanatory inscriptions were needed. With this clue, the analysis of the other drawings should be a relatively simple undertaking. It is readily seen that four of the drawings of Ambleteuse and Boulogne (nos. 16–17, 19–20) fall into a group because of obvious similarities in both drafting technique and handwriting. Close scrutiny of the handwriting assures us that the inscriptions were all written by the same hand. Perhaps the easiest way to perceive this is by comparing the phrase 'The ynche stondeth for C fote' (or its equivalent) that indicates the scale on each of the drawings. But it is also apparent that we have here a different hand from that of Clerks A and B. We are led then to compare the handwriting on the drawings with that of the clerk who wrote one of the four extant letters from Rogers (no. 2), and to whom we may henceforth refer as Clerk C. The general impression is that the two hands are the same; comparison of the majuscules *D*, *R*, *S*, and *T* and of the minuscules *b*, *e*, *g*, and *h* confirms this impression. If further proof is needed, the formation of several words that are found in the letter and on one or more of the drawings—such as 'and', 'boloigne', 'haven', and 'the'—leaves the fact practically beyond doubt that Clerk C wrote the inscriptions on these four drawings.

There still remain to be considered the inscriptions on the other drawings attributable to Rogers. It is barely possible that the inscriptions on the plat of the Guines town walls and on the 'Country of Guynes and Bolenois' (nos. 11 and 15) were made by Rogers himself, although to maintain this claim one would have to assume that Rogers used a neater and more formal hand than is to be found in the proven samples of his writing. But the likelihood of this possibility is diminished by the character of another inscription which was surely written by Rogers himself. On Plat A of Hull Manor there is inscribed in the centre of the floor-plan a brief phrase, 'voyd groũde' (no. 14). Every letter in both words can be matched exactly with the same formations in Rogers's holograph. This inscription, when combined with Rogers's endorsement of the 'Country of Guynes and Bolenois' (no. 9), suggests that whenever he did pen an endorsement or an inscription on a drawing, he used exactly the same hand that he did when writing letters or signing official documents. Since this hand was a readable cursive, but not a formal Secretary, the conclusion that it was the only hand which he employed supports our earlier hypothesis that whenever Rogers wished to have inscriptions placed on one of his formal drawings, he used a trained clerk to write those inscriptions in a Secretary Hand which he himself did not possess.

Regarding Plat C of Ambleteuse (no. 18) little need be said, for it is obvious that the two short inscriptions on the plat constitute forms of handwriting later than the time of Rogers. It may be noted, however, that the phrase 'For Hambletow' was written by the same hand that wrote the inscription 'Country of Guynes and Bolenois' on the plat that we have been referring to by that name.

We may now summarize our conclusions relative to the table of handwriting samples given at the beginning of this Appendix. Regarding Rogers's own handwriting, we have one letter entirely written by him (no. 4), his signature on six other letters (nos. 1–3, 5–7), a receipt of his account as Clerk of Ordnance (no. 10), an endorsement of one plat (no. 9), a brief inscription on another plat (no. 14), and an outside possibility of inscriptions on two others (nos. 11 and 15). From the pen of Clerk A there is the account of the Clerk of Ordnance (no. 10) and the inscriptions on two drawings relating to the work at Hull (nos. 12 and 13). To Clerk B's hand there may be ascribed two letters (nos. 1 and 3), the endorsement of the plan of Hull Manor (no. 8), and the receipt of account of the Clerk of Ordnance (no. 10). From Clerk C there is one letter (no. 2) and the inscriptions on four drawings relating to Boulogne and Ambleteuse (nos. 16–17, 19–20).

These conclusions contain significant implications for this study of the drawings associated with Rogers's career. The important differences between the various samples of handwriting found on the drawings and in the documents seem to weaken the conclusion that Rogers executed those drawings which we have attributed to him on documentary, circumstantial, and stylistic evidence. But this objection can be overcome by supposing that Rogers employed his clerks to inscribe those explanatory words, phrases, and notes which he felt the drawings needed. That this was the case has been demonstrated by identifying the various hands in the inscriptions with those clerks who worked for Rogers and who produced for him documents that are still available for comparison with the plats. Finally, by showing not only that clerks inscribed the plats, but that in particular it was Rogers's clerks who inscribed them, we have made the handwriting evidence, negative as it appears at first glance, actually substantiate the claim that John Rogers was the draughtsman of the plats in question.

APPENDIX B

Leonard Beckwith's Survey of Hull Manor, 1538

(Hull, Corporation Archives, M 112)

THERE is in the sayd market gate (Lowgate) a mansion place being the pryncypall hows called the Hull Garthe with diverse buylding upon it. That is to say in the ffirst entre into the same mansion place oute of the saide street called market gate. Ther is first a gatehows buylded with timber and covered with tile wherin is two chambers and the entrye buylded on bothe sydes with warehouses of tymbre werk covered with tile conteyning in length an hundred fote and in brede xx fote. Also there is a towre entringe into the court three stories of heyght buylded with bryck and stone and covered with lead wherin be three chambers conteyning every chambre xviii fete square with chymneis and Jak' to every one of them.

Also the court within the said towre conteyneth half an acre of ground paved with stone and buylded rounde abowte with xvii chambres of bryk and stone and tyled. Wherof vii chambers is benethe and x above with chymneis & Jakys to them. Also the Hall conteyneth in lenght lx fote and in brede xl fote buylded with bryk and stone and coveryd with tyle. And at the West ende of the same hall is a great chambre conteyning in length lx fote and in brede xx fote buylded with bryk and also at the est ende of the hall is the buttery pantrey and a chamber above them tyled. Also there is next them a great kychen xx fote square covered above with lead and the larder and the skolery covered with tyle. Also there is in the North side of the hall a Chapell conteyning in length xxviii fote and in brede xv fote made of bryk and stone and covered with lead with a closett and sight into the Chapell. And under the same is two chambres wyth a chymney. Also on the North Syde of the said court there is a ponde yarde with a dove house in yt conteyning an acre of grownd walled rounde abowte with bryk.

Also in the West Side of the said court there is a closse conteyning two acres of pasture walled rounde abowte with bryk. Also there is another gardyne adioyning to the same with a dove house in it conteyning half an acre walled. Also there is on the South Syde of the hall another court conteyning a rode buylded rounde abowte with stone and bryk covered with tyle with bakehows, brewhouses & other howses & lodgyng.

BIBLIOGRAPHY

PRIMARY SOURCES IN MANUSCRIPT

Hull. Corporation Archives, M 112. Exemplification of Leonard Beckwith's Survey of Hull Manor, 18 August 1538.

London. British Museum.

Add. MS. 5753. Original Documents Relating to Affairs of France, the Low Countries, Germany, etc. XVI & XVII Cent.

Add. MS. 6176. Miscellaneous Collections.

Add. MS. 46348.

Cott. MSS., Aug. I. (Various maps and plats.)

Harl. MSS. 1647 and 1648. Accounts of works at Sandgate Castle.

Harl. MS. 1724. Works Accounts at Wark Castle.

Royal MS. 7. C. XVI. Household Accounts of Henry VIII, etc.

London. Public Record Office.

E. 36/235–47. Works Accounts of Hampton Court Palace.

E. 101/173/7. Particulars of the Account of Richard de Eccleshale, Treasurer of Calais [1356–8].

E. 101/205/6. Account Book relating to Guines [1527–9].

E. 101/205/9–10, 13–14. Controlment Books of works in the Marches of Calais [1539–42].

E. 101/206/2, 6, 9–10, 12–14. Account Books relating to works in Marches of Calais [1536–42].

E. 101/207/1. An Account relating to Boulogne [1544–6].

E. 101/207/3–4. Account Books of expenses of surveying the king's possessions of Boulogne and Newhaven [1548–50].

E. 101/540/4. Account of works at Guines Castle [1535–6].

E. 101/540/7. Duplicate of the Account of Sir Richard Cotton, Treasurer of Boulogne and Newhaven [1548–50].

S.P. 1/1–246. State Papers, Henry VIII. General Series.

S.P. 10/1–19. State Papers, Domestic, Edward VI.

S.P. 68/1–15. State Papers, Foreign, Edward VI.

Maidstone. Kent Archives Office.

P.R.C. (Probate Records Canterbury Diocese), A. Vol. 32, fol. 30. Will of John Rogers.

Sa/Ac3. Sandwich Old Red Book, 1527–51.

PRIMARY SOURCES IN PRINT

BEAUCHESNE, JOHN DE, AND BAILDON, JOHN. *A Booke Containing Divers Sortes of Hands.* London, 1571.

The Book of Architecture of John Thorpe in Sir John Soane's Museum. Edited, with Biographical and Analytical Studies by John Summerson. Walpole Society, vol. xl. Glasgow, 1966.

Buck, Samuel and Nathaniel. [*A Collection of Engravings of Castles, Abbeys, and Towns in England and Wales.*] London, 1726–52. 5 vols.

The Chronicle of Calais, in the Reigns of Henry VII. and Henry VIII. to the Year 1540. Ed. John G. Nichols. Camden Society, no. 35. London, 1846.

Commentaires de Blaise de Monluc, Maréchal de France. Ed. Paul Courteault. 'Collection de textes pour servir à l'étude et à l'enseignement de l'histoire.' Paris, 1911–25. 3 vols.

A Commentary of the Services and Charges of William Lord Grey of Wilton. Ed. Sir Philip de M. G. Egerton. Camden Society, no. 40. London, 1847.

De L'Orme, Philibert. *Le Premier Tome de l'architecture de Philibert de l'Orme.* Paris, 1568.

Dürer, Albrecht. *Etliche underricht zu befestigung der Stett Schloss und Flecken.* Nürenberg, 1527.

Fer, Nicholas de. *Introduction à la fortification. Troisième partie. Suitte des forces de l'Europe.* Paris, 1693.

Hall, Edward. *The Union of the two noble and illustrate Famelies of Lancastre & Yorke.* London, 1548.

Haigneré, D. 'Boulogne sous l'occupation anglaise en 1549', *Bulletin de la société académique de l'arrondissement de Boulogne*, i (1864–72), 432–45.

—— 'Rapport d'un espion français en 1549', ibid. iv (1885–90), 274–81.

Hamy, E. T. 'Extraits de la correspondance de Maugiron relatifs à l'expulsion des Anglais du Boulonnais', ibid. vii (1904–7), 618–35.

Historical Manuscripts Commission. *Calendar of the Manuscripts of the Most Hon. The Marquis of Salisbury, K.G., Preserved at Hatfield House, Hertfordshire.* Part ii. London, 1888.

—— *The Manuscripts of the Right Honourable F. J. Saville Foljambe, of Osberton.* London, 1897.

Lambert d'Ardres. *Historia Comitum Ghisnensium.* Ed. Joh. Heller. *Monumenta Germaniae Historica, Scriptores*, xxiv. Hanover, 1879.

Leland, John. *The Itinerary of John Leland in or about the Years 1535–1543.* Ed. Lucy T. Smith. London, 1907–10. 5 vols.

Le Patourel, John, ed. *The Building of Castle Cornet, Guernsey.* I, *Documents Relating to the Tudor Reconstruction.* [Guernsey], 1958.

Madden, Frederick. *Catalogue of the Manuscript Maps, Charts, and Plans, and of the Topographical Drawings in the British Museum.* London, 1844. 2 vols.

Ortelius, Abraham. *Theatrum Orbis Terrarum. Antwerp, 1570.* Introduction by R. A. Skelton. *Theatrum Orbis Terrarum*, First Series, vol. iii. Amsterdam, 1964.

Public Record Office. *Acts of the Privy Council of England.* New Series. Ed. John R. Dasent. London, 1890–1907. 32 vols.

—— *Calendar of State Papers, Domestic Series, of the Reigns of Edward VI, Mary, Elizabeth.* London, 1856–72. 12 vols.

—— *Calendar of State Papers, Foreign Series, of the Reign of Edward VI, 1547–1553.* London, 1861.

—— *Calendar of State Papers, Foreign Series, of the Reign of Mary, 1553–1558.* London, 1861.

—— *Calendar of the Carew Manuscripts Preserved in the Archiepiscopal Library at Lambeth.* London, 1867–73. 6 vols.

—— *Calendar of the Patent Rolls Preserved in the Public Record Office. Edward II.* London, 1894–1904. 5 vols.

—— *Calendar of the Patent Rolls Preserved in the Public Record Office. Edward III.* London, 1891–1916. 16 vols.

PUBLIC RECORDS OFFICE. *Calendar of the Patent Rolls Preserved in the Public Record Office. Richard II.* London, 1895–1909. 6 vols.

—— *Calendar of the Patent Rolls Preserved in the Public Record Office. Edward VI.* London, 1924–9. 6 vols.

—— *Calendar of the Patent Rolls Preserved in the Public Record Office. Philip and Mary.* London, 1936–9. 4 vols.

—— *Calendar of the State Papers Relating to Ireland, of the Reigns of Henry VIII, Edward VI, Mary, and Elizabeth.* London, 1860–1912. 11 vols.

—— *Letters and Papers, Foreign and Domestic, of the Reign of Henry VIII.* London, 1862–1932. 23 vols. in 38.

ROSNY, A. DE. *Album historique du Boulonnais.* Neuville-sous-Montreuil, 1892.

—— 'Documents inédits ou rarissimes, concernant les sièges de Boulogne, 1544–1549', *Mémoires de la société académique de l'arrondissement de Boulogne-sur-Mer*, xxvii (1912), 379–540.

RYMER, THOMAS. *Foedera, Conventiones, Literae et Cujuscunque Generis Acta Publica.* 2nd ed., London, 1727–35. 20 vols.

SELVE, ODET DE. *Correspondance politique de Odet de Selve, ambassadeur de France en Angleterre (1546–1549).* Ed. G. Lefèvre-Pontalis. Paris, 1888.

SPEED, JOHN. *The Theatre of Great Britaine: Presenting an Exact Geography of the Kingdoms of England, Scotland, Ireland, and the Iles adioyning.* London, 1611.

STANEWELL, L. M., ed. *Calendar of the Ancient Deeds, Letters, Miscellaneous Old Documents, &c., in the Archives of the Corporation.* Hull, 1951.

SECONDARY WORKS

ACKERMAN, JAMES S. 'Architectural Practice in the Italian Renaissance', *Journal of the Society of Architectural Historians*, xiii, no. 3 (1954), 3–11.

BAGWELL, RICHARD. *Ireland under the Tudors, with a Succinct Account of the Earlier Period.* London, 1963. 3 vols.

BERTRAND, P. J. B. *Précis de l'histoire physique, civile et politique de la ville de Boulogne-sur-Mer.* Boulogne, 1828–9. 2 vols.

BIDDLE, MARTIN. 'Nonsuch Palace', *History Today*, xi (1961), 206–14.

—— 'Nonsuch 1959–60: An Interim Report', *Surrey Archaeological Collections*, lviii (1961), 1–20.

BIRD, JAMES. *The Major Seaports of the United Kingdom.* London, 1963.

BLASHILL, THOMAS. *Evidences relating to the Eastern Part of the City of Kingston-upon-Hull.* Hull, 1903.

BLOMFIELD, REGINALD. *Architectural Drawing and Draughtsmen.* London, 1912.

BLUNT, ANTHONY. *Philibert de l'Orme.* London, 1958.

BOYS, WILLIAM. *Collections for an History of Sandwich in Kent.* Canterbury, 1792.

BRIGGS, MARTIN S. *The Architect in History.* Oxford, 1927.

BROWN, LLOYD. *The Story of Maps.* Boston, 1949.

BURROWS, MONTAGU, *Cinque Ports*, 2nd ed. 'Historic Towns.' London, 1888.

CARDEN, R. W. 'The Italian Artists in England during the Sixteenth Century', *Proceedings of the Society of Antiquaries of London*, 2nd ser., xxxiv (1911–12), 171–204.

COCKLE, MAURICE J. D. *A Bibliography of English Military Books up to 1642 and of Contemporary Foreign Works.* London, 1900.

COLVIN, H. M. *Biographical Dictionary of English Architects, 1660–1840*. Cambridge, Mass., 1954.

COLVIN, H. M., ed. *The History of the King's Works*. Vols. i and ii. *The Middle Ages*, by R. Allen Brown, H. M. Colvin, and A. J. Taylor. London, 1963.

DE LA CROIX, HORST. 'The Literature on Fortification in Renaissance Italy', *Technology and Culture*, iv (1963), 30–50.

—— 'Military Architecture and the Radial City Plan in Sixteenth-Century Italy', *Art Bulletin*, xlii (1960), 263–90.

DENT, JOHN. *The Quest for Nonsuch*. London, 1962.

Dictionary of National Biography. Edited by Sir Leslie Stephen and Sir Sidney Lee. London, 1949–50. 22 vols.

DILLON, HAROLD A. 'Calais and the Pale', *Archaeologia*, liii, pt. 2 (1893), 289–388.

An English Garner: Tudor Tracts, 1532–1588. Introd. by A. F. Pollard. New York, 1903.

ENLART, CAMILLE. 'Boulogne monumental', *Congrès Archéologique de France* (Arras, 1880), 291–310.

—— 'Les monuments anciens de Boulogne', *Boulogne-sur-Mer et la région boulonnaise*. 2 vols. (Boulogne-sur-Mer, 1899), i. 153–300.

—— *Manuel d'archéologie française depuis les mérovingiens jusqu'à la renaissance. Deuxième partie: architecture civile et militaire*. 2nd ed. rev. et augmentée. Paris, 1929–32. 2 vols.

FINCH, JAMES K. *The Story of Engineering*. Garden City, N.Y., 1960.

FORDHAM, SIR HERBERT G. *Some Notable Surveyors and Map-Makers of the Sixteenth, Seventeenth, and Eighteenth Centuries and their Work*. Cambridge, 1929.

FRANCE. Commission departementale des monuments historiques du Pas-de-Calais. *Dictionnaire historique et archéologique du Pas-de-Calais*. Arras, 1875–84.

FRANKL, PAUL. *The Gothic*. Princeton, 1960.

FROST, CHARLES. *Notices Relative to the Early History of the Town and Port of Hull*. London, 1927.

GARDINER, DOROTHY. *Historic Haven: The Story of Sandwich*. Derby, England, 1954.

GENT, THOMAS. *Annales Regioduni Hullini: or, The History of the Royal and Beautiful Town of Kingston-upon-Hull*. London, 1735.

GERARD, H. 'Ambleteuse; un procès où se mêle l'histoire', *Annales Boulogne-sur-Mer*, ii (1904), 189 ff.

GILLE, BERTRAND. *Les ingénieurs de la Renaissance*. Paris, 1964.

GIROUARD, MARK. 'Three Gothic Drawings in the Smithson Collection', *Journal of the Royal Institute of British Architects*, 3rd ser., lxiv, no. 1 (1956), 35–36.

HADLEY, GEORGE. *A New and Complete History of the Town and County of the Town of Kingston-upon-Hull*. Kingston-upon-Hull, 1788.

HAIGNERÉ, D. 'Dictionnaire topographique de la France comprenant les noms de lieux anciens et modernes, arrondissement de Boulogne-sur-Mer', *Mémoires de la société académique de l'arrondissement de Boulogne-sur-Mer*, xi (1882), 1–392.

HALE, JOHN R. *The Art of War and Renaissance England*. 'Folger Booklets on Tudor and Stuart Civilization.' Washington, 1961.

—— 'The Early Development of the Bastion: An Italian Chronology *c*. 1450–*c*. 1534', *Europe in the Late Middle Ages*, ed. J. R. Hale, J. R. L. Highfield, and B. Smalley (London, 1965), 466–94.

HARVEY, A. S. *The de la Pole Family of Kingston upon Hull*. Hull, 1957.

HARVEY, JOHN H. *An Introduction to Tudor Architecture*. London, 1949.

HARVEY, JOHN H. 'Early Tudor Draughtsmen', *The Connoisseur Coronation Book* (London, 1953), 97–102.
—— *English Mediaeval Architects: A Biographical Dictionary down to 1550.* London, 1954.
—— *Henry Yevele,* c. *1320 to 1400: The Life of an English Architect.* London, 1946.
HECTOR, L. C. *The Handwriting of English Documents.* London, 1958.
HÉLIOT, PIERRE. *Le château de Boulogne-sur-Mer.* Paris, 1933.
—— 'Les fortifications de Boulogne sous l'occupation anglaise (1544–1550)', *Revue du Nord*, xl (1958), 5–38.
HIRST, JOSEPH H. 'The Castle of Kingston-upon-Hull', *Transactions of the East Riding Antiquarian Society*, iii (1895), 24–39.
HOGG, O. F. G. *The Royal Arsenal: Its Background, Origin, and Subsequent History.* London, 1963. 2 vols.
HOPE, W. H. ST. J. *Windsor Castle.* London, 1913. 2 vols.
JENKINS, FRANK. *Architect and Patron: A Survey of Professional Relations and Practice in England from the Sixteenth Century to the Present Day.* London, 1961.
JENKINSON, HILARY. *The Later Court Hands in England from the Fifteenth to the Seventeenth Century.* Cambridge, 1927.
KNOOP, DOUGLAS, and JONES, G. P. 'The Decline of the Mason Architect in England', *Journal of the Royal Institute of British Architects*, 3rd ser., xliv (1937), 1004–7.
LANGENSKIÖLD, ERIC. *Michele Sanmicheli: The Architect of Verona, His Life and Works.* Uppsala, 1938.
LAW, ERNEST. *The History of Hampton Court Palace.* London, 1885–91. 3 vols.
LEFEBVRE, ABBÉ. *Histoire générale et particulière de la ville de Calais, et du Calaisis.* Paris, 1766. 2 vols.
LENNEL, F. *Histoire de Calais.* Calais, 1909–11. 2 vols.
LEWIS, MICHAEL. *Armada Guns: A Comparative Study of English and Spanish Armaments.* London, 1961.
LYNAM, EDWARD. *British Maps and Map-Makers.* London, 1944.
—— 'The Development of Symbols, Lettering, Ornament and Colour on English Maps', *Proceedings of the British Records Association*, no. 4 (1938), 20–34.
—— 'English Maps and Map-Makers of the Sixteenth Century', *Geographical Journal*, cxvi (1950), 7–28.
—— *The Mapmaker's Art: Essays on the History of Maps.* London, 1953.
LYON, JOHN. *The History of the Town and Port of Dover and of Dover Castle.* Dover, 1813. 2 vols.
MACDONALD, ALEC. 'Plans of Dover Harbour in the Sixteenth Century', *Archaeologia Cantiana*, xlix (1938), 108–26.
MACIVOR, IAIN. 'The Elizabethan Fortifications of Berwick-upon-Tweed', *Antiquaries Journal*, xlv (1965), 64–96.
MACKIE, J. D. *The Earlier Tudors, 1485–1558.* Oxford, 1952.
MAGGIOROTTI, LEONE A. *Architetti e architettura militari.* In *L'Opera del genio italiano all' estero.* Series 4: *Gli architetti militari.* Rome, 1933–9. 3 vols.
MINET, WILLIAM. 'Some Unpublished Plans of Dover Harbour', *Archaeologia*, lxxii (1922), 185–224.
MORGAN, F. W. *Ports and Harbours.* London, 1952.
OMAN, CHARLES. *A History of the Art of War in the Sixteenth Century.* London, 1937.

O'NEIL, B. H. ST. J. *Castles and Cannon: A Study of Early Artillery Fortifications in England.* Oxford, 1960.

—— 'Stefan von Haschenperg, an Engineer to King Henry VIII, and his work', *Archaeologia*, xci (1945), 137–55.

PARSONS, W. B. *Engineers and Engineering in the Renaissance.* Baltimore, 1939.

POLLARD, A. F. *The History of England from the Accession of Edward VI to the Death of Elizabeth (1547–1603).* London, 1929.

POPHAM, A. E. 'Hans Holbein's Italian Contemporaries in England', *Burlington Magazine*, lxxxiv (1944), 12–17.

PORTER, WHITWORTH. *History of the Corps of Royal Engineers.* London, 1889. 2 vols.

POUNCEY, PHILIP. 'Girolamo da Treviso in the Service of Henry VIII', *Burlington Magazine*, xcv (1953), 208–11.

RIGAUX, M. E. 'Notes sur les noms des rues de la Basse Boulogne aux XVIe et XVIIe siècles', *Bulletin de la société académique de l'arrondissement de Boulogne*, vi (1900–3), 90–130.

ROBINSON, A. H. W. *Marine Cartography in Britain: A History of the Sea Chart to 1855.* Leicester, 1962.

ROSNY, J. HECTOR DE. *Histoire de Boulonnais.* Amiens, 1868–73. 4 vols.

RUTTON, W. L. 'Henry VIII's Castles at Sandown, Deal, Walmer, Sandgate, and Camber', *Archaeologia Cantiana*, xxiii (1898), 24–30.

—— 'Sandgate Castle, A.D. 1539–40', *Archaeologia Cantiana*, xx (1893), 228–57.

SALZMAN, LOUIS F. *Building in England down to 1540: A Documentary History.* Oxford, 1952.

SHEAHAN, JAMES J. *History of the Town and Port of Kingston-upon Hull.* 2nd ed. Beverley, 1866.

SHELBY, L. R. 'The Role of the Master Mason in Mediaeval English Building', *Speculum* xxxix (1964), 387–403.

SHEPPARD, THOMAS. *The Evolution of Kingston-upon-Hull as shown by its Plans.* Hull, 1911.

STEINBERG, S. H. 'Medieval Writing-Masters', *The Library*, 4th ser., xxii (1941), 1–24.

SUMMERSON, JOHN. *Architecture in Britain, 1530 to 1830.* 'Pelican History of Art', no. Z3. 4th ed. London, 1963.

TANNENBAUM, SAMUEL A. *The Handwriting of the Renaissance.* New York, 1930.

TICKELL, JOHN. *The History of the Town and County of Kingston upon Hull.* Hull, 1798.

TOOLEY, R. V. *Maps and Map-Makers.* 2nd ed. London, 1952.

—— 'Maps in Italian Atlases of the Sixteenth Century, Being a Comparative List of the Italian Maps Issued by Lafreri, Forlani, Duchetti, Bertelli and Others, Found in Atlases', *Imago Mundi*, iii (1939), 12–44.

TOY, SIDNEY. *A History of Fortification from 3000 B.C. to A.D. 1700.* New York, 1955.

TRAVIS-COOK, JOHN. *Notes Relative to the Manor of Myton.* Hull, 1890.

TURNER, T. HUDSON, and PARKER, J. H. *Some Account of Domestic Architecture in England from the Conquest to Henry VIII.* Oxford, 1851–9. 3 vols.

VAILLANT, V. J. 'L'Advis de la qualité de ceulx qui sont dedans Boulogne (1559)', *Bulletin de la société de l'arrondissement de Boulogne*, vii (1904–7), 334–55.

—— 'La tour d'Ordre et son nom gaulois', *Notes boulonnaises, variétés* (Boulogne-sur-Mer, 1889), 93–113.

WEBB, HENRY J. *Elizabethan Military Science: The Books and the Practice.* Madison, 1965.

WILLIAMS, IOLA A. *Early English Watercolours and Some Cognate Drawings by Artists born not later than 1785.* London, 1952.

WOOD, MARGARET. *The English Mediaeval House.* London, 1965.

NOTES ON THE PLATES

1. 'Plat A of Guines.' B.M., Cott. MSS., Aug. I, Supp. 14. (101×71·5 cm.; slightly irregular.) Paper. Sepia ink lines; no wash. *Inscriptions*: Thest gatte / pyrtons / The keipe / Whettellis / The Inshe conteynyth L fotte / The West gatte. *By courtesy of the Trustees of the British Museum.*

2. 'Plat B of Guines.' B.M., Cott. MSS., Aug. I. ii. 51. (98·5×67 cm.; irregular.) Parchment. Black ink lines; sepia ink used in construction lines, gold ink in border scale; grey wash in outside rim of walls. No endorsements or inscriptions. *By courtesy of the Trustees of the British Museum.*

3. 'Plat C of Guines.' B.M., Cott. MSS., Aug. I, Supp. 2. (127×139·5 cm.; irregular.) Paper. Sepia ink lines; no wash. No endorsements or inscriptions. *By courtesy of the Trustees of the British Museum.*

4. 'Plat D of Guines.' B.M., Cott. MSS., Aug. I. ii. 12. (64·3×39·7 cm.) Paper. Sepia ink lines; sepia wash in windows of castle. No endorsements or inscriptions. *By courtesy of the Trustees of the British Museum.*

5. 'The Plot of Guins.' B.M., Cott. MSS., Aug. I. ii. 52. (113×80 cm.) Paper. Sepia and black ink lines; blue wash for moats; red wash for castle walls; green wash for grass. Endorsement: THE PLOT OF GUINS. *Inscription*: A hundrid fote betwixt the dove and the travers wall. *By courtesy of the Trustees of the British Museum.*

6. Bird's-eye view of Hull. B.M., Cott. MSS., Aug. I. i. 83. (77·4×56·6 cm.) Paper. Sepia and black ink lines; sepia wash for land, pink wash for town walls and housetops, greyish-green wash for trees and gardens. *Inscriptions*: BEVERLAY / COTYNGAM / BARTON / HVL / The Charter-house. *By courtesy of the Trustees of the British Museum.*

7. 'Kyngeston vpon Hvll' by W. Hollar. B.M., Dept. of Prints and Drawings, Hollar, vol. x, p. 23. (30·8×25·8 cm.) *By courtesy of the Trustees of the British Museum.*

8. Plan of Hull fortifications. B.M., Cott. MSS., Aug. I, Supp. 4. (202×69 cm.; irregular.) Paper. Sepia and black ink lines; no wash. *Inscriptions*: The Southende / The lenght betwene the Bulwarke next the Humber and the Castell is mlxvi fote / The whole length of the woorke is mm^{tt} viii^{C} iiii^{xx} xiiii fote / The breadthe of haven at the full see marke is CCxlvi fote. *By courtesy of the Trustees of the British Museum.*

9. Survey of Hull area. B.M., Cott. MSS., Aug. I, Supp. 20. (99×132 cm.) Paper. Sepia and black ink lines; light green wash outlining rivers, roads, and drainage ditches; black wash in Humber. *Inscriptions*: Barton / Skytter mylnes / Humber / Pawle / Headon / Preston / / The Clyffe by hasell / Hasell / The See dyke betwene Hasell and Hull / / The Wolde / The Wolde/ Anlobye / Elley / Overton / darningham welles / darningham sewer / Holdenprice / Cotyngham / This from darningham welles to Hull / This brooke runneth

from darningham Sewer to Stone ferye / The Brooke that springeth at Holdenprice and runneth to Stonefery / The ryver that runneth from Cotingham to Stone ferye / The waye to Cotingham / The waye to Mewlande / The waye to Beverlaye / The Chartrehouse / Cowscates / Mewelande / Stone ferye / Hull ryver / Sutton / Holdernesse / Scala miliaria. *By courtesy of the Trustees of the British Museum.*

10. 'Plat A of Hull Manor.' B.M., Cott. MSS., Aug. I, Supp. I. (122×127·5 cm.; irregular.) Paper. Sepia to blackish ink lines; no wash. *Inscription*: voyd groũde. *By courtesy of the Trustees of the British Museum.*

11. 'Plat B of Hull Manor.' B.M., Cott. MSS., Aug. I. ii. 11. (41·5×26 cm.) Paper. Sepia ink lines; sepia wash in windows and on alternating steps in stairs. No endorsements or inscriptions. *By courtesy of the Trustees of the British Museum.*

12. 'Plat C of Hull Manor.' B.M., Cott. MSS., Aug. I. i. 84. (52·7×32·5 cm.) Paper. Sepia to blackish ink lines; no wash. *Endorsement* (on dorse): A new plat made by the same Rogers, of the king his hyghnis manno^{r} of Hulle, the xxvth of June. *By courtesy of the Trustees of the British Museum.*

13. 'Plat D of Hull Manor.' B.M., Cott. MSS., Aug. I. ii. 13. (76·5×78·3 cm.; irregular.) Paper. Sepia ink lines; sepia wash in some doors and windows. No endorsements or inscriptions. *By courtesy of the Trustees of the British Museum.*

14. 'Country of Guynes and Bolenois.' B.M., Cott. MSS., Aug. I. ii. 75. (54·2×35·8 cm.) Parchment. Sepia ink lines; sepia wash in rivers and outlining some hills. *Endorsements*: Country of Guynes and Bolenois; (on dorse) A plat of the contre of gy[nes and] bolinnoys. *Inscriptions*: Mentory / Cressynere / Redlington / Bukkold / Campe / Ardes / balingham / Guynes / Hamis / Cales / Sandingfeld / Scales clyffe / / ffynes mylle / Wyer / Margyson / Sula Molyne / Beverkyn / the m[ar]she / Whytsand / Oltersey / blaknes / / bell castell / boleyn brygge / Waymylle / Boleyn / / Cowting / Botes / davern / The hilles / Samber boyes / The hilles / heding Abbe / quese / pont brigge / Saint Etyne / hardeloo / le fforte / / the ynche contayneth a myle. *By courtesy of the Trustees of the British Museum.*

15. 'Plat A of Ambleteuse.' B.M., Cott. MSS., Aug. I. i. 59. (54·7×79 cm.) Paper. Sepia and black ink lines; sepia wash faintly outlining shores. *Inscriptions*: Ambletw / Depe / Depe / Depe / Depe / Depe / the haven mowthe / / The Sande hylles / The naturall pere / The shore towardes the blaknesse / The ynche stondeth for C fote / the lowe water mark / The See / / The Sande hylles / The Shore towardes boloigne / The See. *By courtesy of the Trustees of the British Museum.*

16. 'Plat B of Ambleteuse.' B.M., Cott. MSS., Aug. I. ii. 68. (55·3×41·7 cm.) Paper. Dark sepia ink lines; sepia wash used for outlining shores and colouring sea; flankers of fort marked with red lines. Monogram in lower right-hand corner: NR. *Inscriptions*: Thear is no higher grownde then the place that the town stondeth on the bulwerke to the londe warde stondeth of lyke heygthe and dyssendyth a lyttle towarde the haven / / Sand hylles / Sand hylles / The shore to the black nesse warde / The ynche conteyneth ii C fote / The lowe water marke / The See / / A hyll / A valley / A lytte valey / A hyll / Ambletewghe / the churche / Lowe Sand hylles / High sand hylles / The shore to boloign

warde / The chanell / the hows on the peere / The haven mowthe / The See. *By courtesy of the Trustees of the British Museum.*

17. 'Plat C of Ambleteuse.' B.M., Cott. MSS., Aug. I. ii. 73. (47·4×36 cm.) Parchment. Black ink lines. Yellow wash outlining hills and fort; brownish wash for ships' hulls and for outlining shores. *Endorsements*: Hauen Etue; For Hambletow. *By courtesy of the Trustees of the British Museum.*

18. 'Plat D of Ambleteuse.' B.M., Cott. MSS., Aug. I. ii. 8. (92·3×66 cm.) Parchment. Black ink lines; bluish-green wash outlining hills; blue wash in moat around fort; ships decorated in red. *Endorsements*: Haven Etewe; haven Etu. *By courtesy of the Trustees of the British Museum.*

19. 'Survey I of Boulogne.' B.M., Cott. MSS., Aug. I, Supp. 5. (155·5×107·5 cm.) Paper. Sepia ink lines; pink wash for Liane River and sea; green wash outlining Fort d'Outreau, the sandhills, and High Boulogne. *Inscriptions*: High bollen / basse bollen / The yownge man / The olde man / The [. . .] of the haven / / Oultre Ewe / . . .er and mar the shipes that ley . . . bollen and such as be w^t^in the havin / / Thise outher thre bowlwarkes be not yet so forwards nor yet so fortyfied as thouther / They wourke now vppon this to fortyfy this as Stronge as the outher / from one bowllwark to thouther is a Cxliii passe / The Sircuit of every bowlwark is CCxlviii passes and this is more [forti]fyd than all the rest / here laith vii batering cannons towards mostreull which prevalith but lyttill / The high Way from the portell to the Seesyde / The Campe of allmens and picards and Divers outher nastions / the highway [to] Mostrewll / The camp of the piemōtois and gascons / Monsg^r^ du bies is camped w[ith]in this churchyard callyd oultre Ewe and is well hallf a myle from the allemans Campe / The Campe of the gentill men and is well a gret myle from thallmens & from the gascons well tou mylle / The Campe of horsmen is well a gret inggliS myle from any outher Compeny & from the fortres. *By courtesy of the Trustees of the British Museum.*

20. 'Survey II of Boulogne.' B.M., Cott. MSS., Aug. I. ii. 77. (95×55 cm.) Paper. Sepia ink lines; sepia wash in streams and for outlining cliffs and shores; red ink dots outlining territorial limits. *Endorsement*: Boleine with the French Fortresse and the Country towards Hardilo. *Inscriptions*: The mownte / highe boloigne / the voide grownde / the towne left owte / the Cytadell / the yonge man / the sand hylles / lowe water mark / The Inche stondeth for v C fote / The See / / The great Medowe / A valey / The ffortresse / The highest grownde / A playne felde sumwhat ffallyng towarde pawlet / The Cleffe / porte hyll / pawlet / The See / lowe water marke / The fote of the hyll / The medowe / A vayley on the hylles side / The medowe / A vayley on the hylle side / The Almayn Camp / A knobbe joyned to the hyll / A valeye / playne grownde / The highe churche / playne / A lyttle hyll / a valey / A hyll / a valey / A hyll / a valley / The Rownde hyll / A lyttle hyll / The stronge valey / A playne felde / A hyll / lowe grownd by ye see / The See / / The Medowe / S. Estyen / The medowe / A valey / A lyttle hyll / playne grownde / A valey / The playne / A hyll / The Clyff / lowe water marke. *By courtesy of the Trustees of the British Museum.*

21. 'Survey III of Boulogne.' B.M., Cott. MSS., Aug. I. ii. 53. (81·5×62·3 cm.) Parchment. Sepia ink lines; orange ink for border scale; green wash outlining shores, hills, and

forts. *Endorsement* (on dorse): of Boullen w^{t} a devyes of a campe for the wynnyng of the frenshe fortyfycac̄on foranenst bullen. *Inscriptions*: The Sande hylles / The See / / The medowe / The camp next the medowe / a valey / a valey / The place to make a mownte / The hyll nexte the churche / a valeye / pawlet hylle / pawlet / The ynch stondethe for v C ffote / The Rode / The Rode / / The medowe / The space betwene the campe / The campe where the Almaynes lay / a valey / The Waye / a valey / The hyll / a valeye / The strayte / A valeye / A mownt naturall / a valey / The camp nexte the see / The stronge valley / The plaine / The See. *By courtesy of the Trustees of the British Museum.*

22. 'Survey IV of Boulogne.' B.M., Cott. MSS., Aug. I. ii. 82. (92 × 58·8 cm.) Paper. Sepia and black ink lines; greenish-blue and brownish wash intermixed throughout plat. *Inscriptions*: Bollen / Old man. *By courtesy of the Trustees of the British Museum.*

23. Map of the Pas de Calais by Nicholas de Nicolay, 1558. From Abram Ortelius's *Theatrum Orbis Terrarum* (1570).

24. An Elizabethan Plat of the Isle of Thanet. B.M., Cott. MSS., Aug. I. i. 54. (42·7 × 44·8 cm.) Paper. Black ink lines; blue wash for water; green outline of seashores and streams; red wash for rooftops. *Endorsement*: Becon / The End of the Bolder / 6000 foote from the sea to this plac / The Newe cute / The Popes head / 4000 / The Gestlyngs / Sandwiche / Sandwiche Mershe / / The Downes / The Brakes / Flattes / The Beare / Flattes / The Channell / Flattes / Litell: Joy / Richborroughe / The Chanell hath gon this Way in old time and may doe / This goeth to Canterbury / / The North Forland / Brodgate brodstaui / Rammys gate / St Lawrens / St. Peters / St. Johns / Margate / Woode Church / Mr. Crips / Burchington / St. Nicholas / Minster / Morton / Sar / The ould chanel but now plainground / Bishop Morton cut this / A dam / A dam / The 3 great dams this way ye water desireth most to goe / Bishop Morton cut this / This chanell ebbith and floweth hither / This Chanell is stopped / Reculvers / The olde mouth / North mouth / The Rode of Gorend / / A B C D E F G H / North / On Mile / half mile. *By courtesy of the Trustees of the British Museum.*

25*a*. Endorsement on Plat C of Hull Manor. (See note on Pl. 12.)

25*b*. Portion of Survey of the Hull area. (See note on Pl. 9.)

25*c*. Portion of Plat A of Hull Manor. (See note on Pl. 10.)

26. Account of the Clerk of Ordnance, 1544. B.M., Add. MS. 5753, fol. 121^{v}. *Text*: at x s the daye. John Rogers—xlii dayes—xxi^{li} / Conductors of ordenaunce xii d Thomas Gore—xlii da'—xlii s / viii d John More—xlii da'—xxviii s / servants at vi d Henry Reade—xlii da'—xxi s / John Gowghe—xlii da'—xxi s / Gregorye Taillo^{r}—xlii da'—xxi s / Philippe Johnes—xlii da'—xxi s / Som' partes—xxviii^{li} xiiii s / / id Sum of thallowaunces—
xx
Cxxiiii^{li} xv^{s} viii^{d} wherof ys to be deducted iiii^{xli} which the same Rogers both receyved in prest And so rem'—xxxiiii^{li} xvi^{s} viii^{d} / Receyvyd by me John Rogers / / Totals—Cxxiiii^{li} xvi^{s} viii^{d}. *By courtesy of the Trustees of the British Museum.*

27. Letter from Rogers to Sir William Paget, 21 February 1547. P.R.O., S.P. 68/13, no. 42. *Text*: After my dowty cōsyderid it may pleas yor mastershipe to vnderstand that within the yonge man I have tracyd or stakyd forthe a little castell and as I thinke very fytt for the same place if it may pleas the kinges m^{ates} owncell that it may be gon in

hand w^{t}hall as m^{r} Dudley is not a little faine therof but tyll I know yor pleasures it shall staye and nothinge be don therto A platt therof have I sent vnto yowe by this bearer every Inche cōtayning x foote thone pointe standythe dyrectely to the hill wheare the windemylle standythe thother pointe to the french forte the gate openithe righte towardes base bolloigne Also m^{r} Dudley hath promisyd that his owne men shall digge the sellers and foundac̄ons and his owne cartes carry the stuffe therto And then no doubte thoughe the platt seme tedious to do yet will it lightely be made and no greate charge neyther shall it lett any other woorks The estate of all the other of the kinges m^{tes} woorkes that I have in hand I wold have declarid but the comissioners bothe first and laste have avewid them with the Cow̃cell heare and advertysyd you therof allredye as they saye whearfore to have troblyd you withall I thought it but superflewus As knowithe the god who ever kepe yor good m^{r}shipe longe to Endure ffrom Bolloigne the xxi of ffebruarye R 1547 / / Yo^{rs} to cōmànd John Rogers

28. Second page of letter from Rogers to the Lord Protector, 27 March 1547. P.R.O. S.P. 68/14, no. 86. *Text*: . . . gyve but I thaught that ix^{s} wer a great price and so sayed that for x^{s} they sholde serve and delyver at boloigne as many as the kynges ma^{te} wolde have Then I prayed hym to sende the colis as was apoynted and as for the price I wolde not meddle that I Referred to yor grace and to the Rest of the kynges ma^{tes} moast honorable cowncell Also the frenche men wolde fayne by coles at boloigne In so moche that at the ffyrste worde they offered xvi^{s} sterlynge ffor a chalder so that I parseave the kynges ma^{te} may be a great gayner yf a Staple of Colis wer hear As tochynge the Reperacyons of howses ffor victualles whiche are styll a doynge and never at an ende and now m^{r} Acher Requyred me to make ii ostes to drye malte on ffor he hathe braught to boloigne muche barley whearvnto belongethe fflowers to make it on so that yf the one be made they wyll call for the other and so shall the charges of the workes apere greater thearbye as it hath Don synce bulaigne hathe byn Englyshe hytherto whearin my Desyer is to knowe yor graces pleasure whether I shall doe it styll or not all other the kynges ma^{te} affares hear as well abowte boloigne as at newe haven goethe Reosably well fforwarde as knowethe god who ever presarve your grace longe to Endewer wrytten at boloign the xxvii of marche your graces allwayes to cōmande John Rogers.

29. Second page of letter from John Rogers to the Privy Council, May 1548. P.R.O., S.P. 68/15, no. 141. *Text*: . . . Reservyd in my handes to be Redy at suche tymes as the paymaster wantyd and so have I yt Ready at altymes to be delyveryd other accomptes have I none to make but my hart ys not alytell hevy that my doynges dothe no better plese yor good lordsheps I havynge no trade nor occupynge in thys wardell but the kynges ma^{tes} bysynes wherin I travell nyght and day and yett so lytell acceptyd as thoughe I ded my own bysynes and lefte the kyngs Alas how lytell Joy brynges thys into my poure harte and how grette hevynesse have I cause to be in to cōsyder all this very moche as knowythe the ever lastynge god who ever preserve yor good lordshepps longe to endure / / yor good lordshepps to cōmande John Rogers.

INDEX

PRINTED IN GREAT BRITAIN
AT THE UNIVERSITY PRESS, OXFORD
BY VIVIAN RIDLER
PRINTER TO THE UNIVERSITY

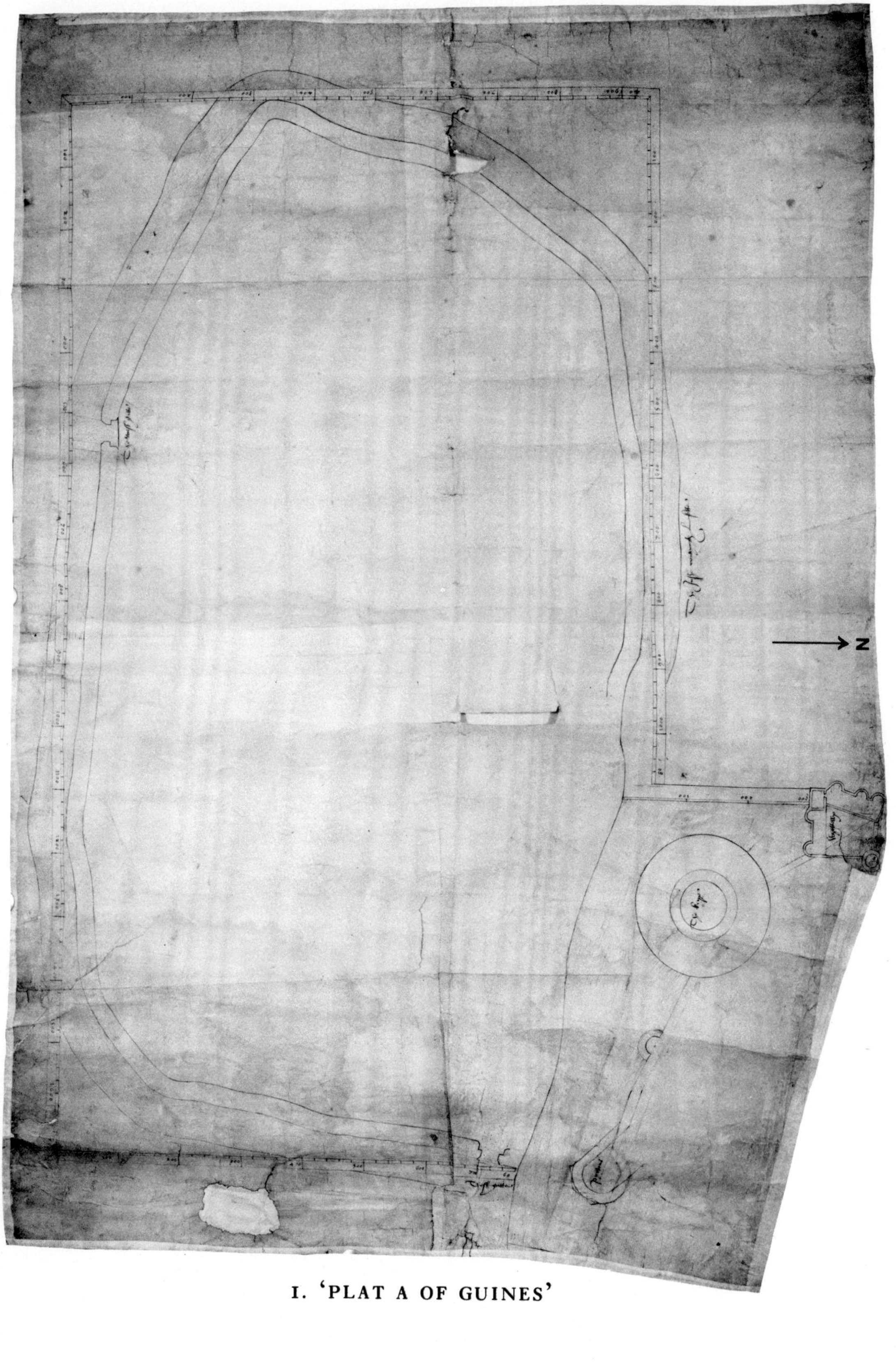

I. 'PLAT A OF GUINES'

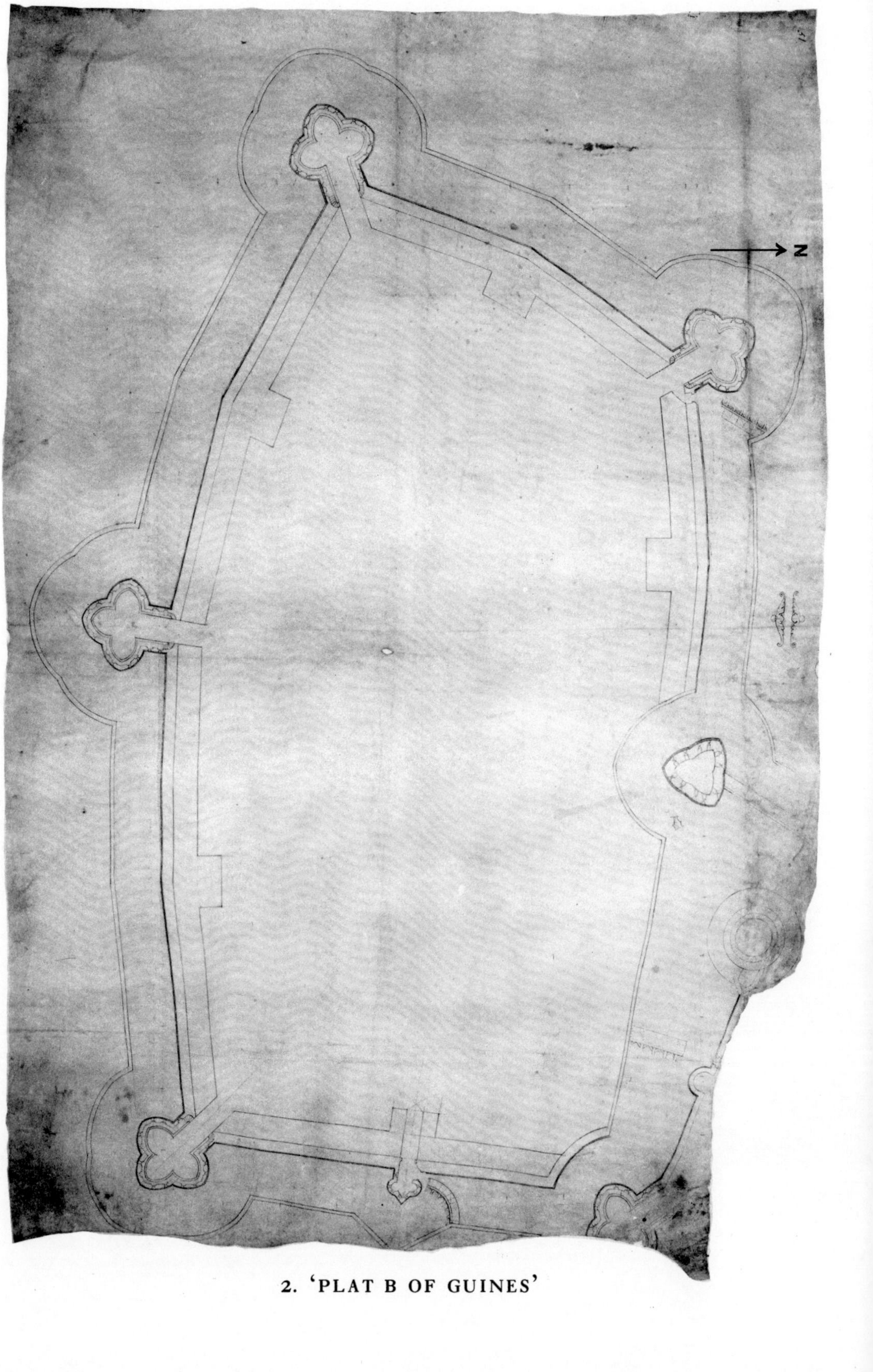

2. 'PLAT B OF GUINES'

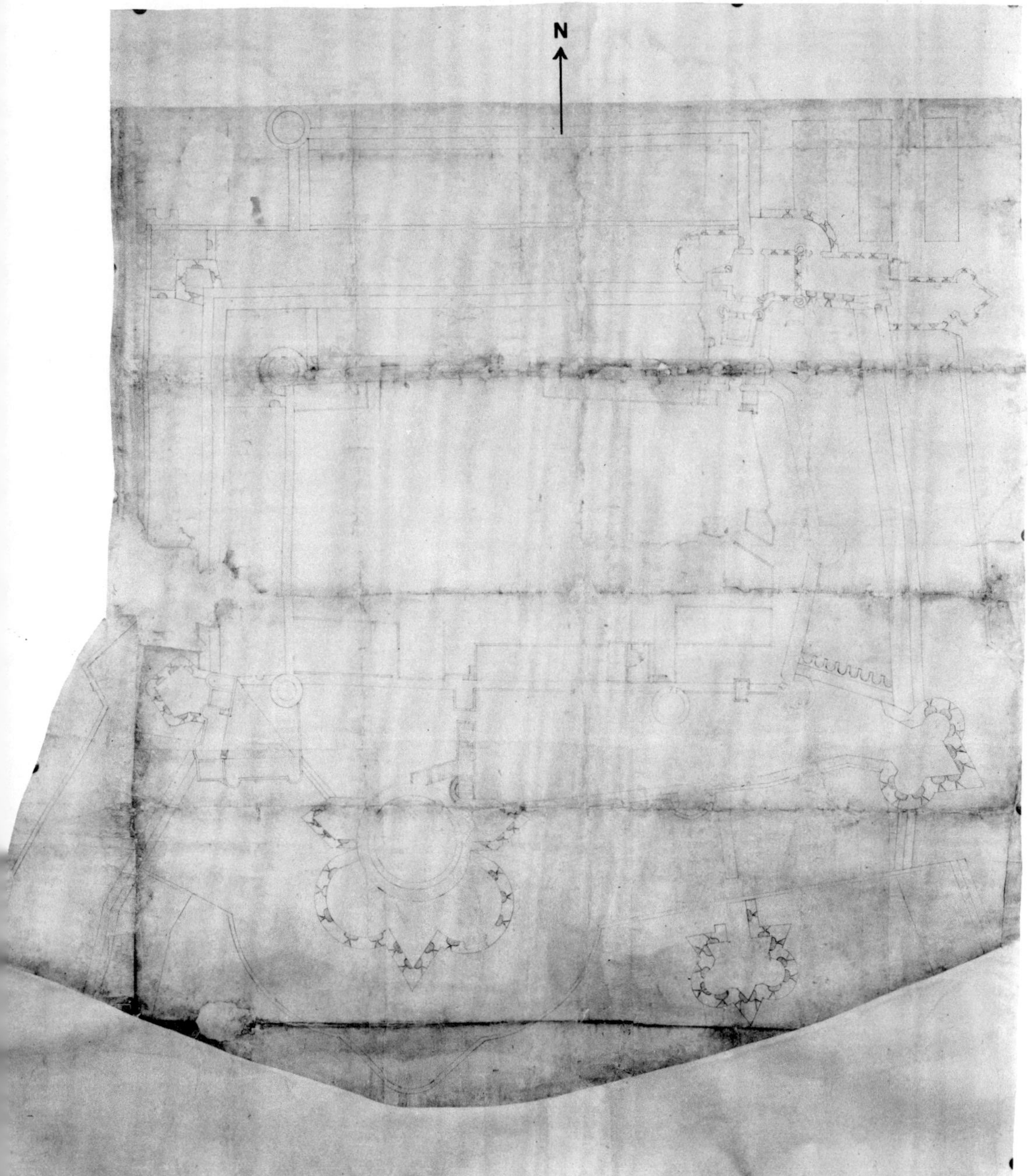

3. 'PLAT C OF GUINES'

4. ‘PLAT D OF GUINES’

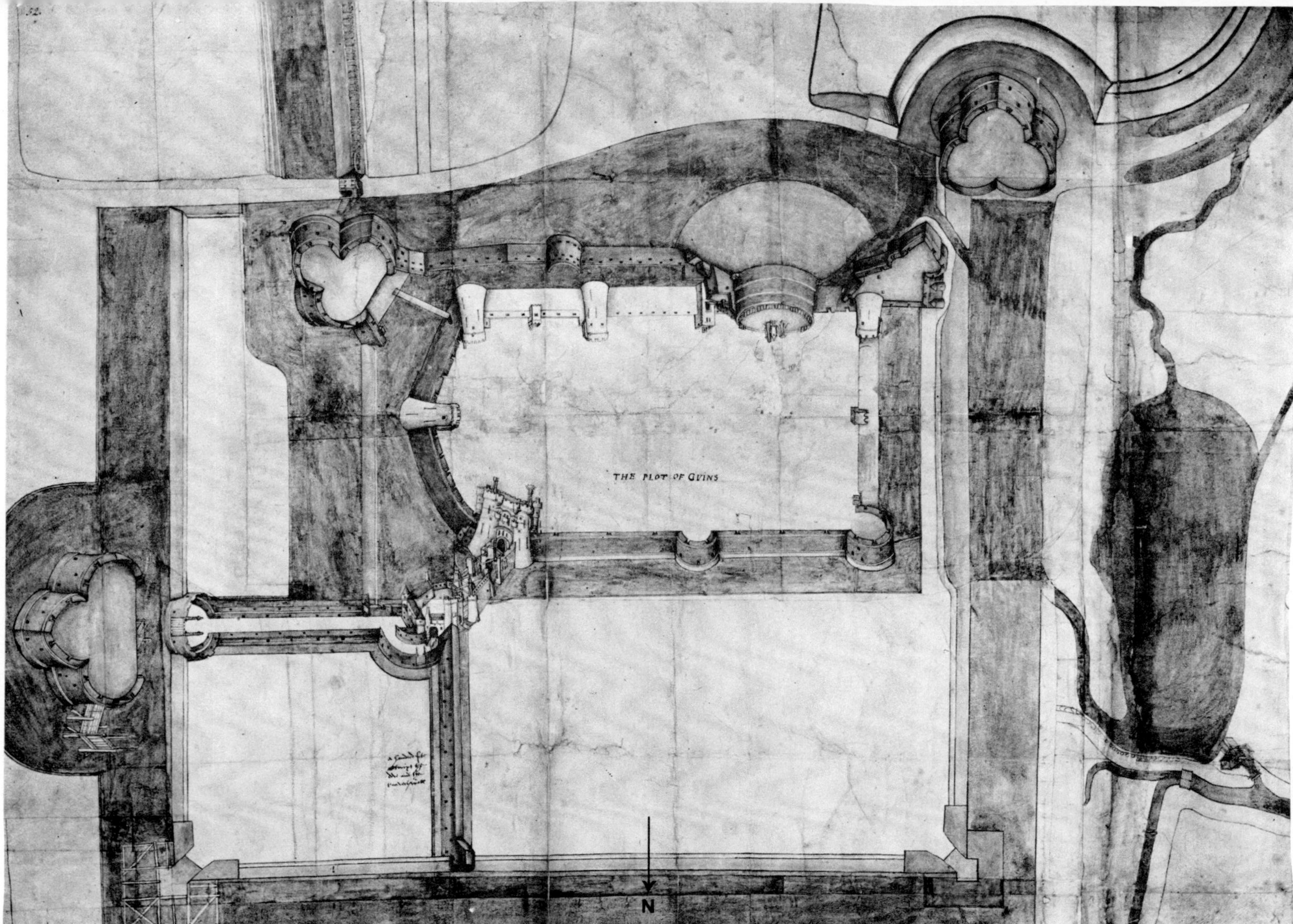

5. 'THE PLOT OF GUINS'

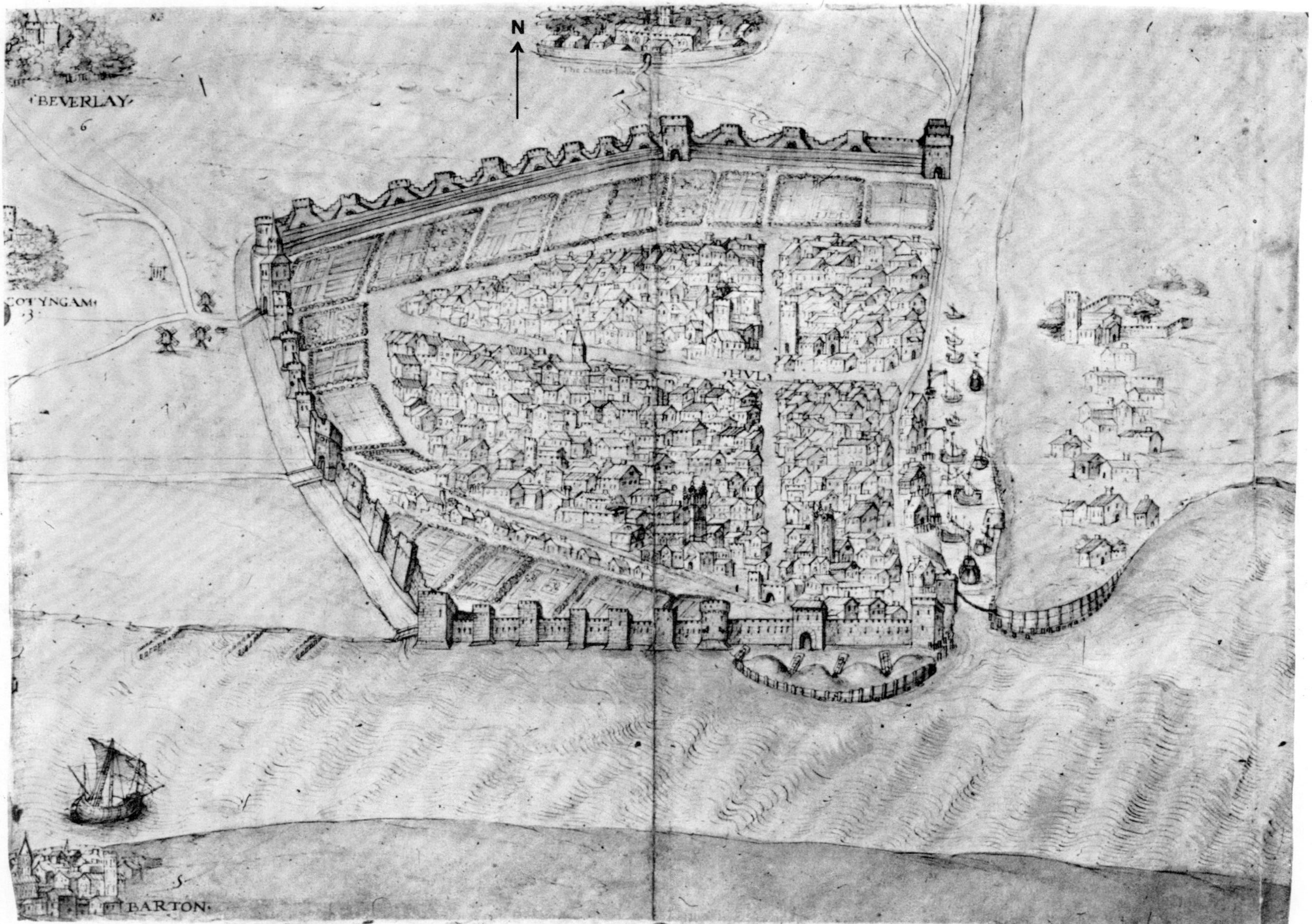

6. BIRD'S-EYE VIEW OF HULL

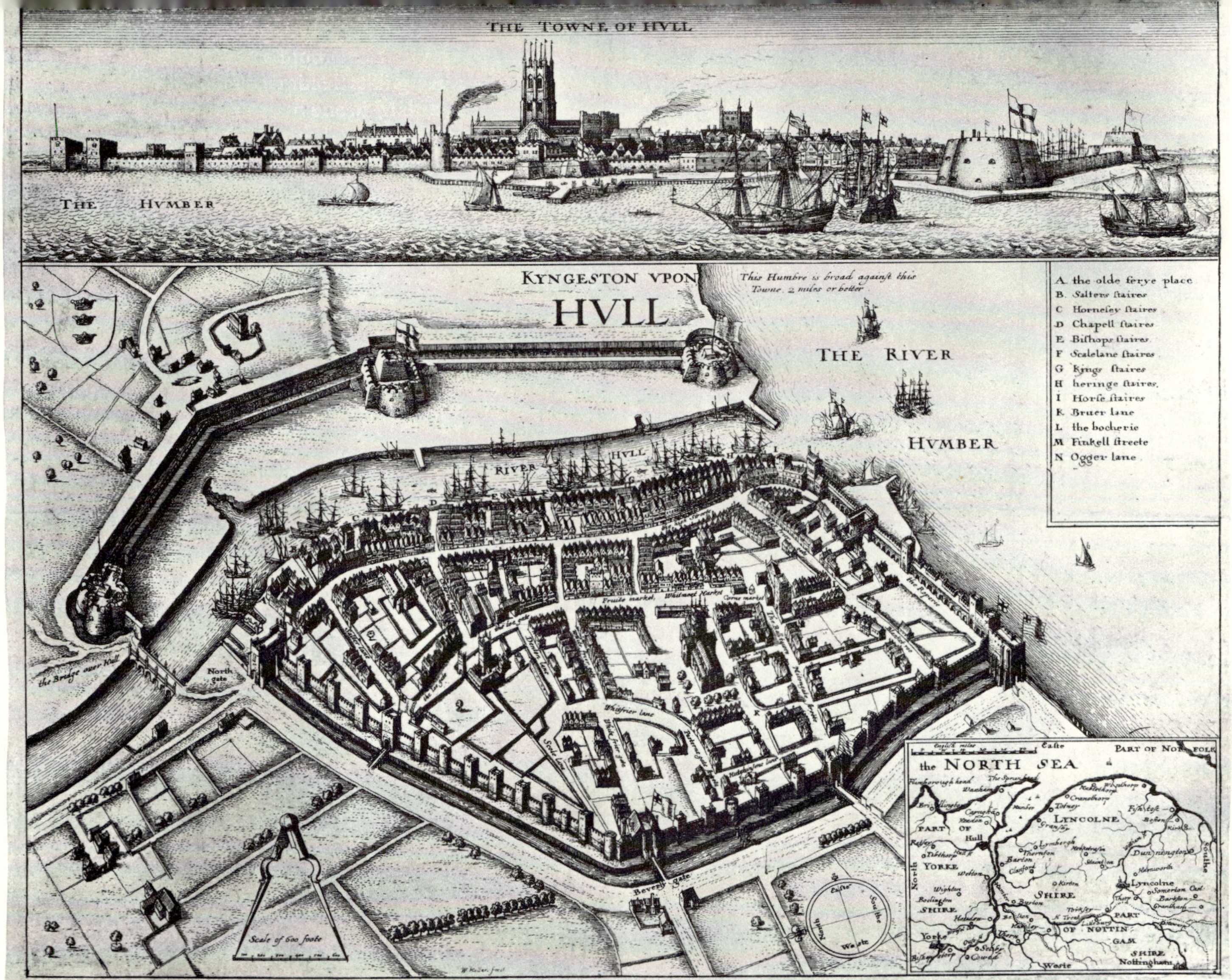

7. 'KYNGESTON VPON HVLL' BY W. HOLLAR

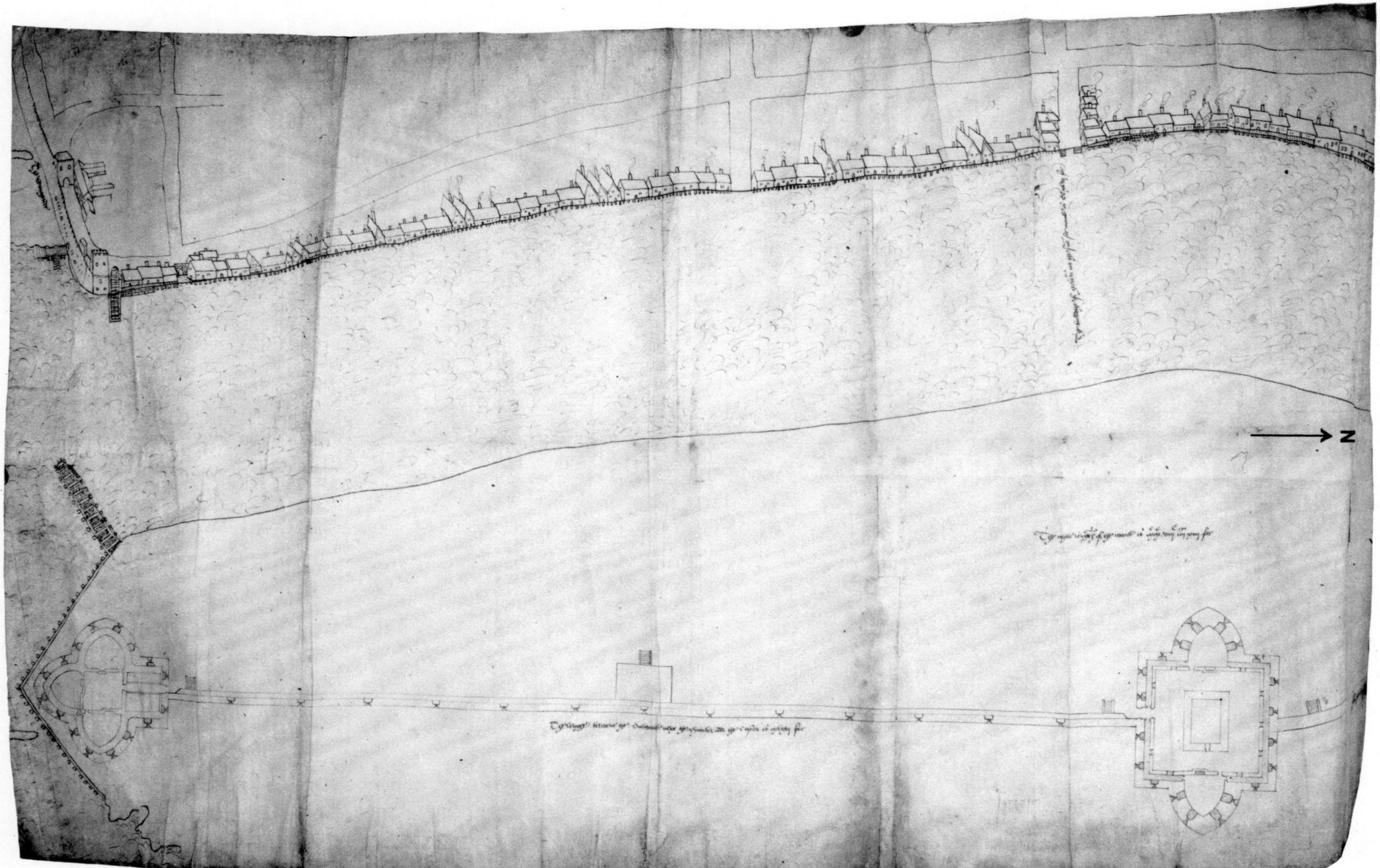

8. PLAN OF HULL FORTIFICATIONS

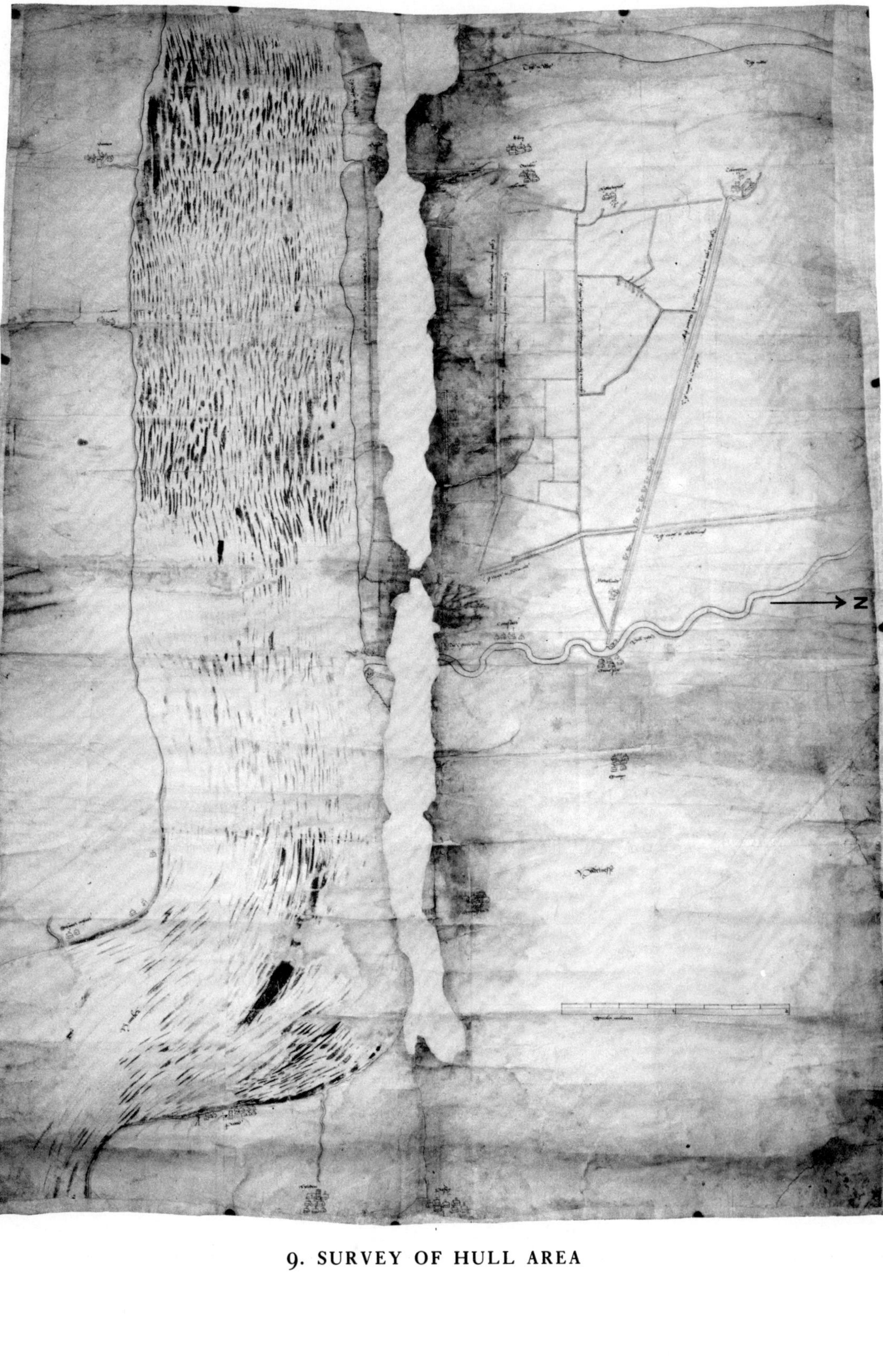

9. SURVEY OF HULL AREA

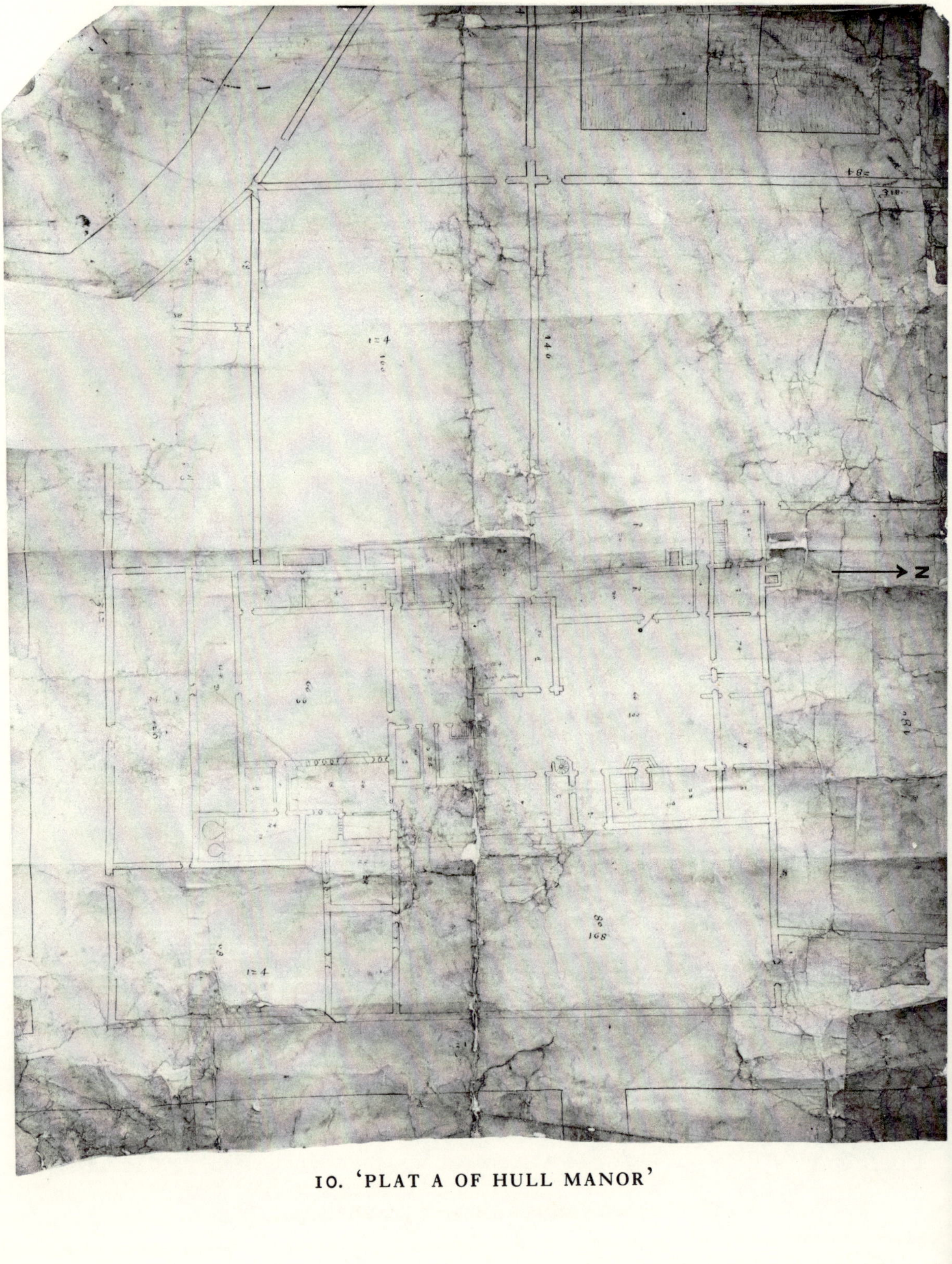

10. 'PLAT A OF HULL MANOR'

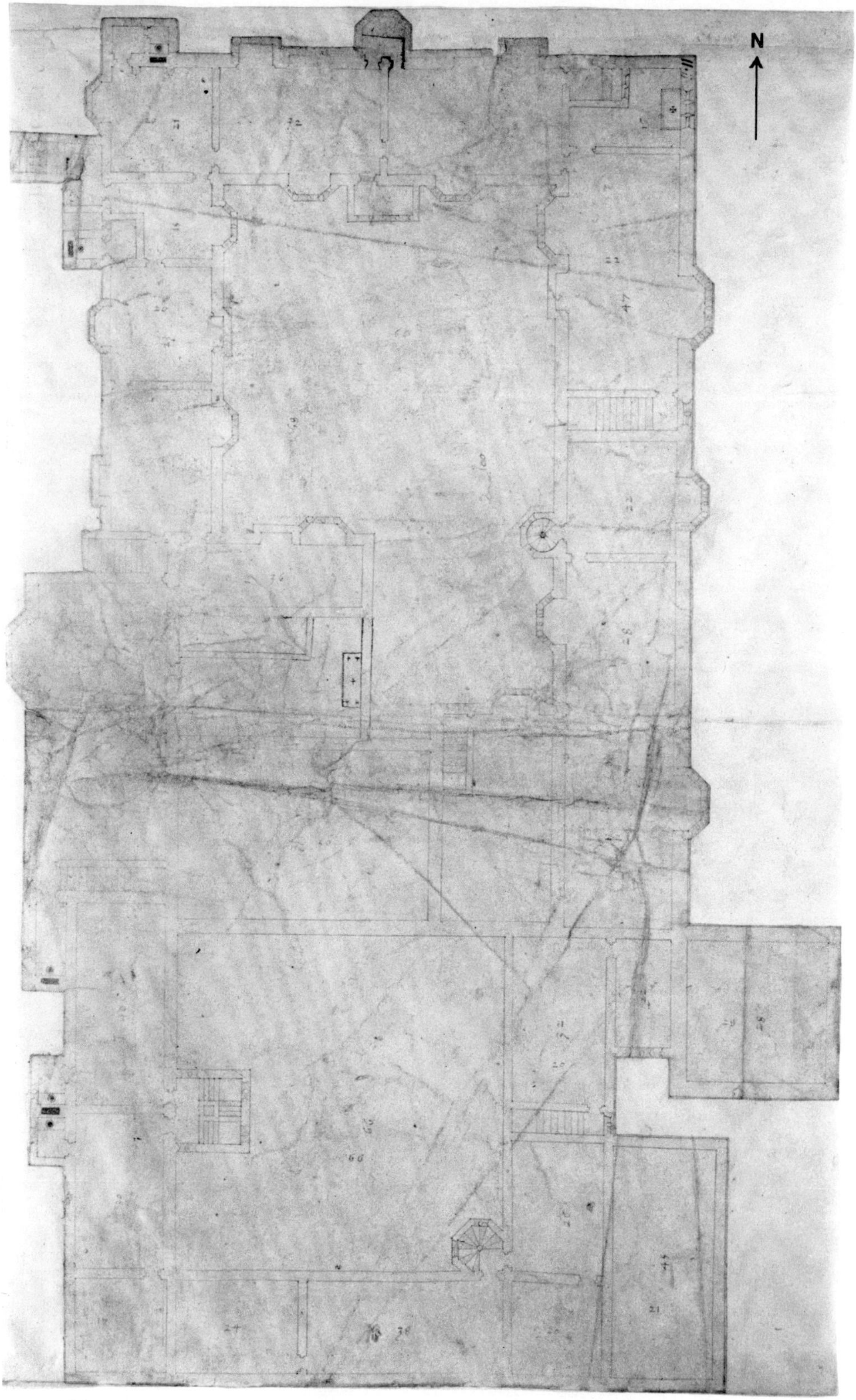

II. 'PLAT B OF HULL MANOR'

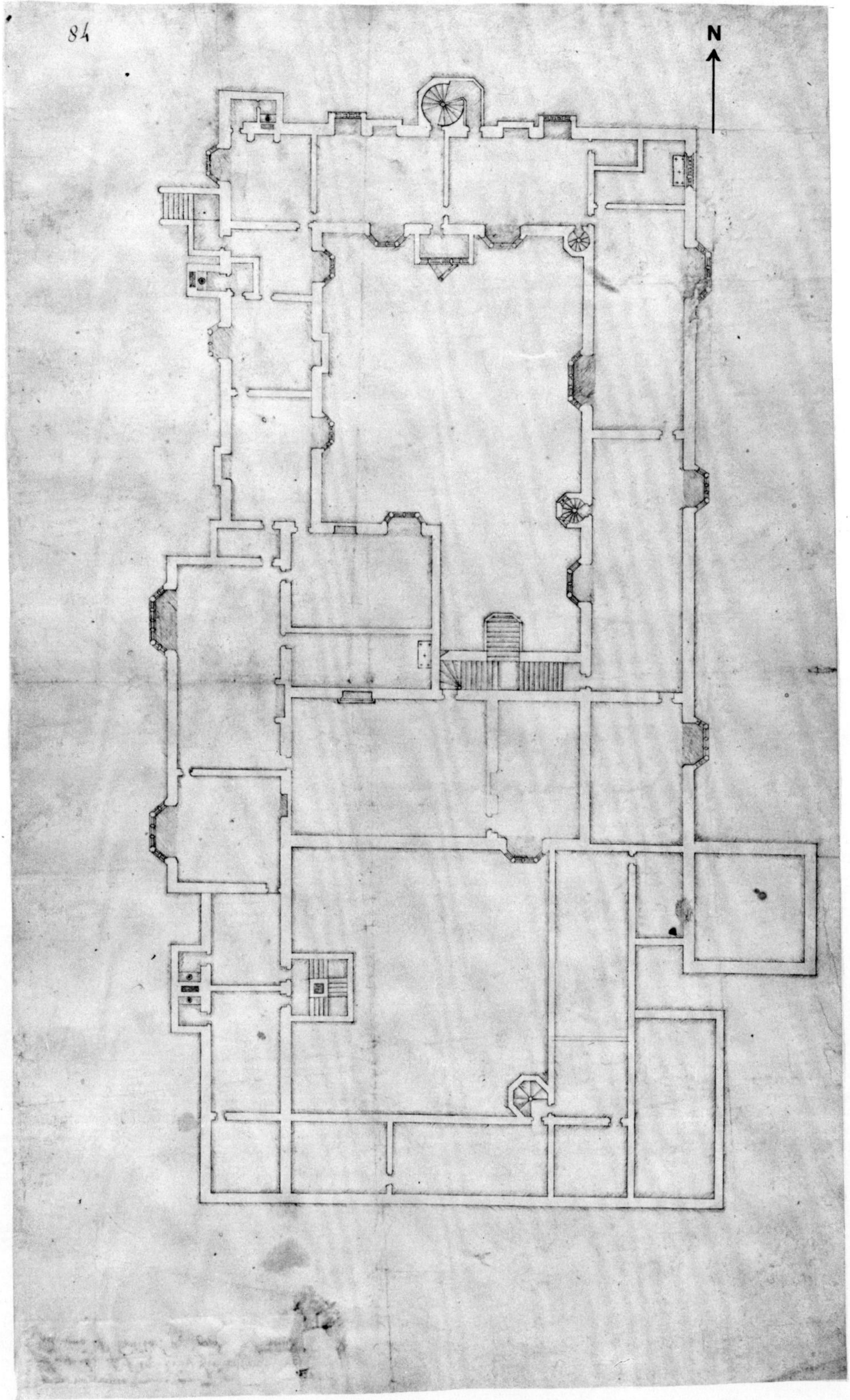

12. 'PLAT C OF HULL MANOR'

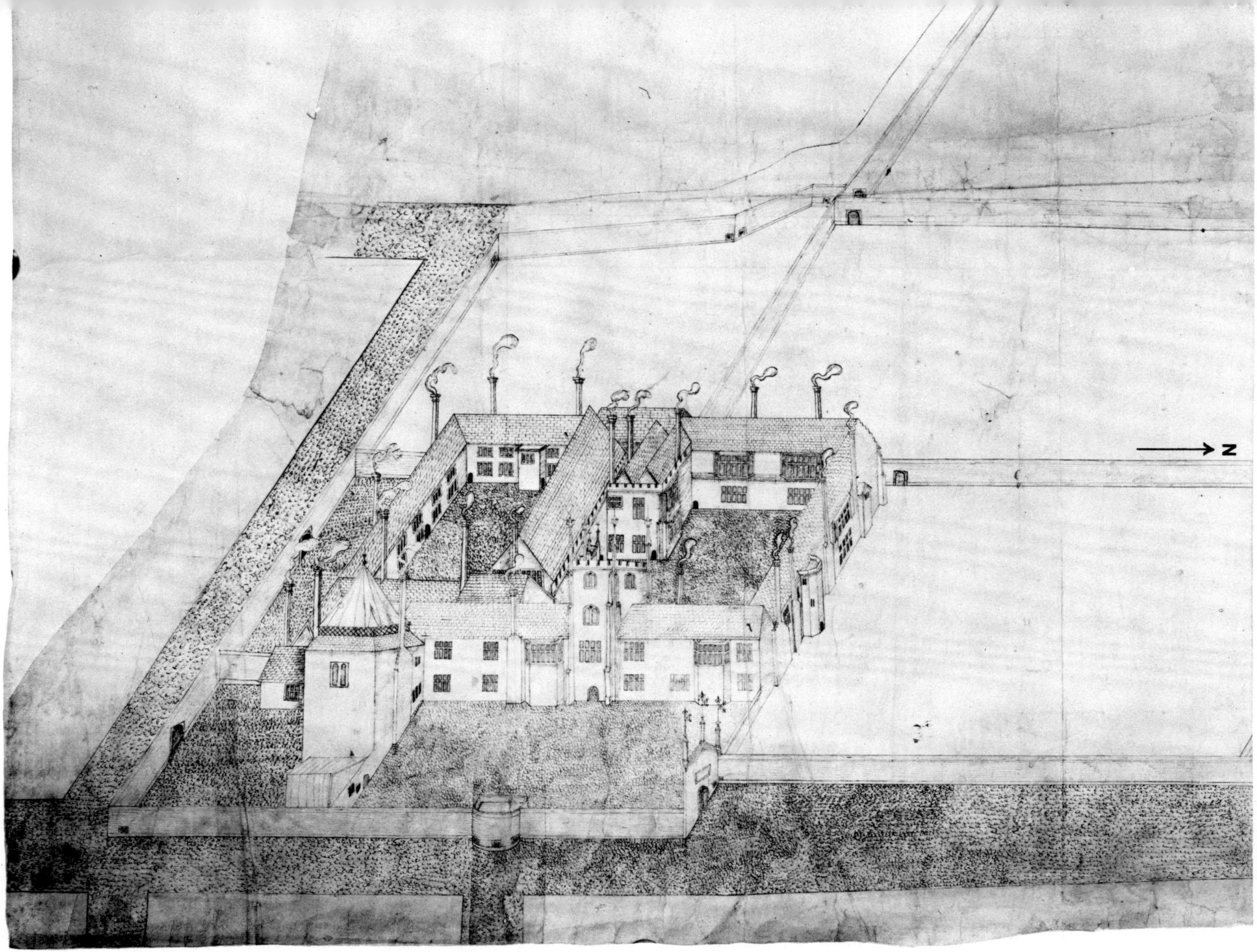

13. 'PLAT D OF HULL MANOR'

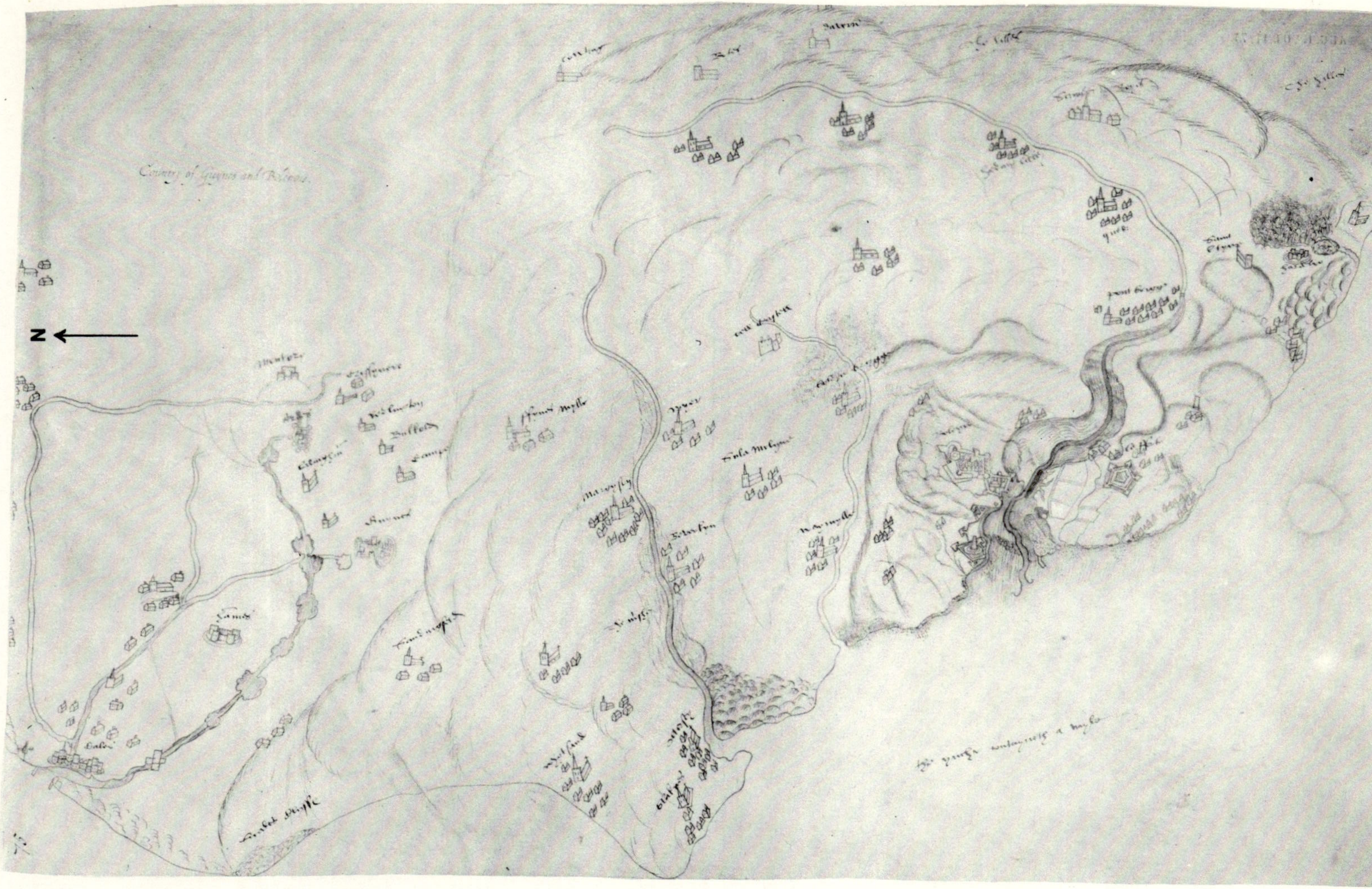

14. 'COUNTRY OF GUYNES AND BOLENOIS'

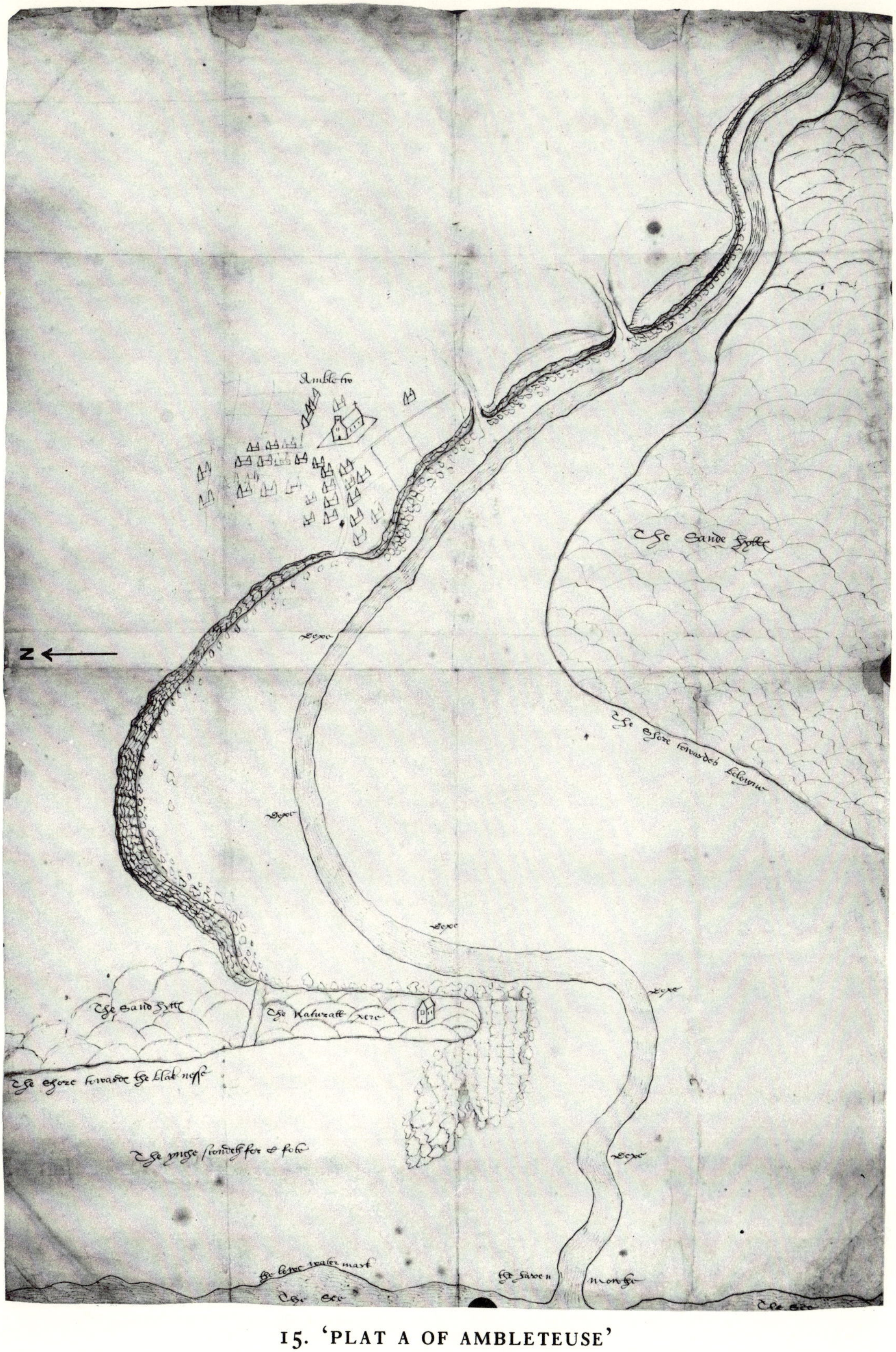

15. 'PLAT A OF AMBLETEUSE'

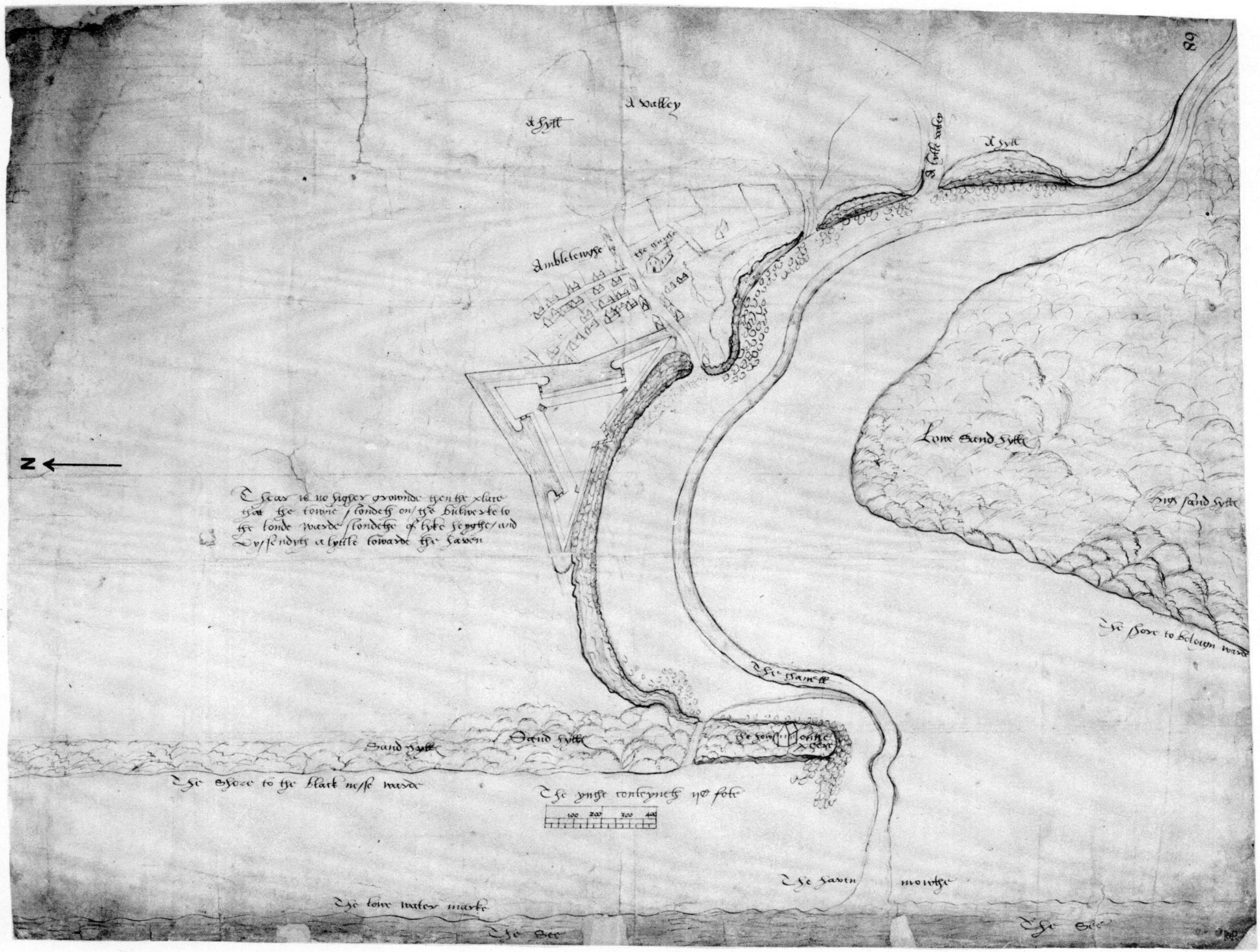

16. 'PLAT B OF AMBLETEUSE'

17. 'PLAT C OF AMBLETEUSE'

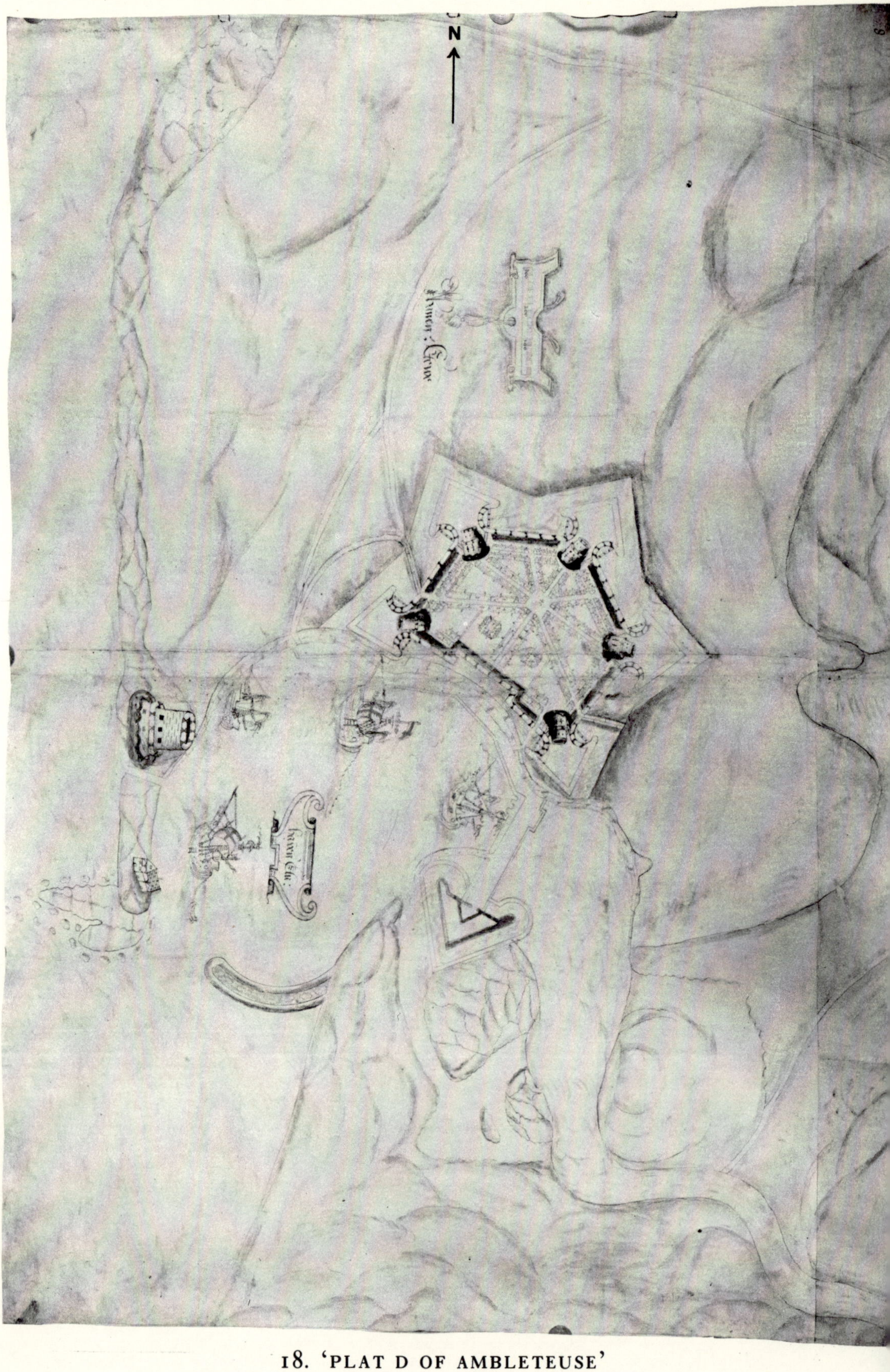

18. 'PLAT D OF AMBLETEUSE'

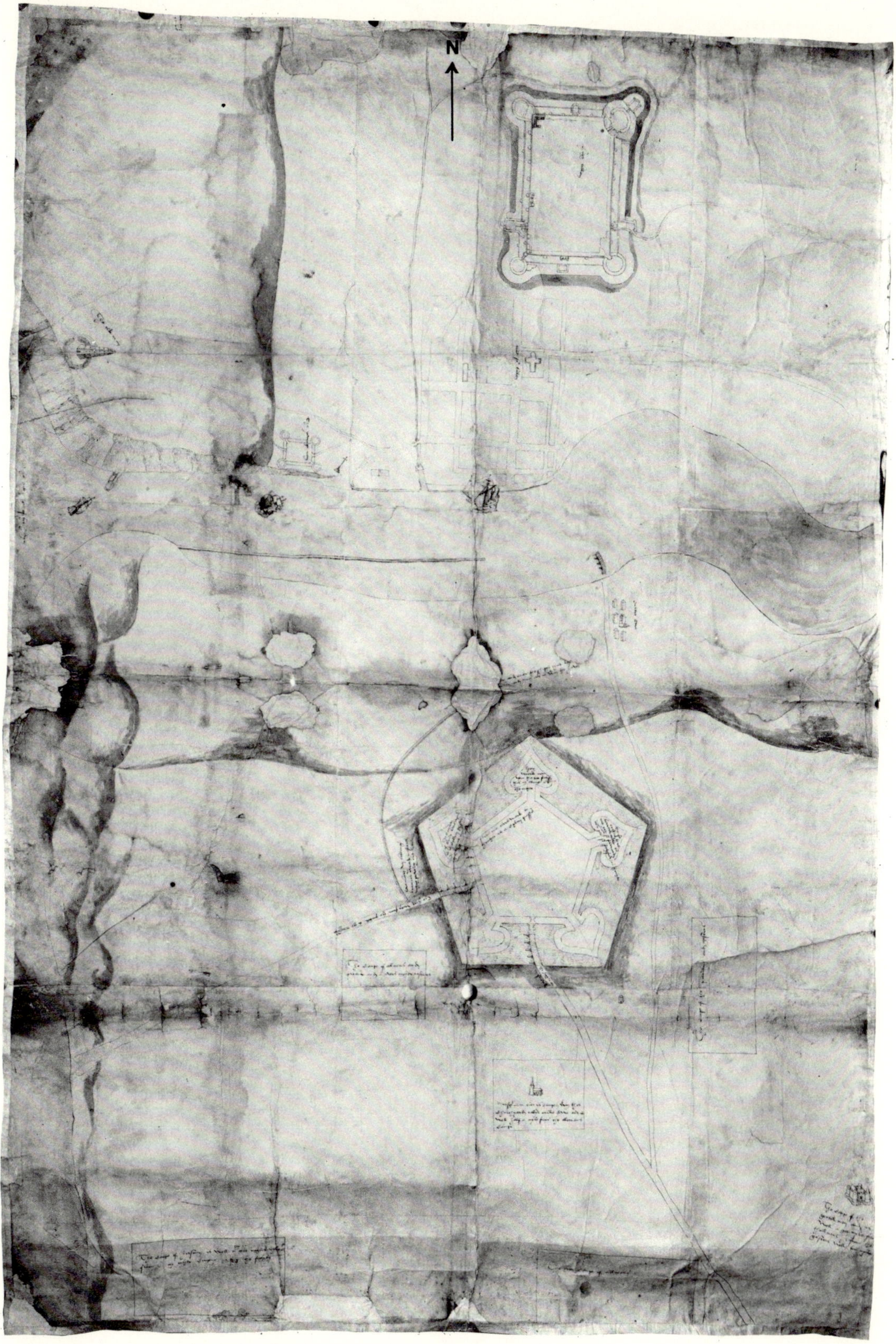

19. 'SURVEY I OF BOULOGNE'

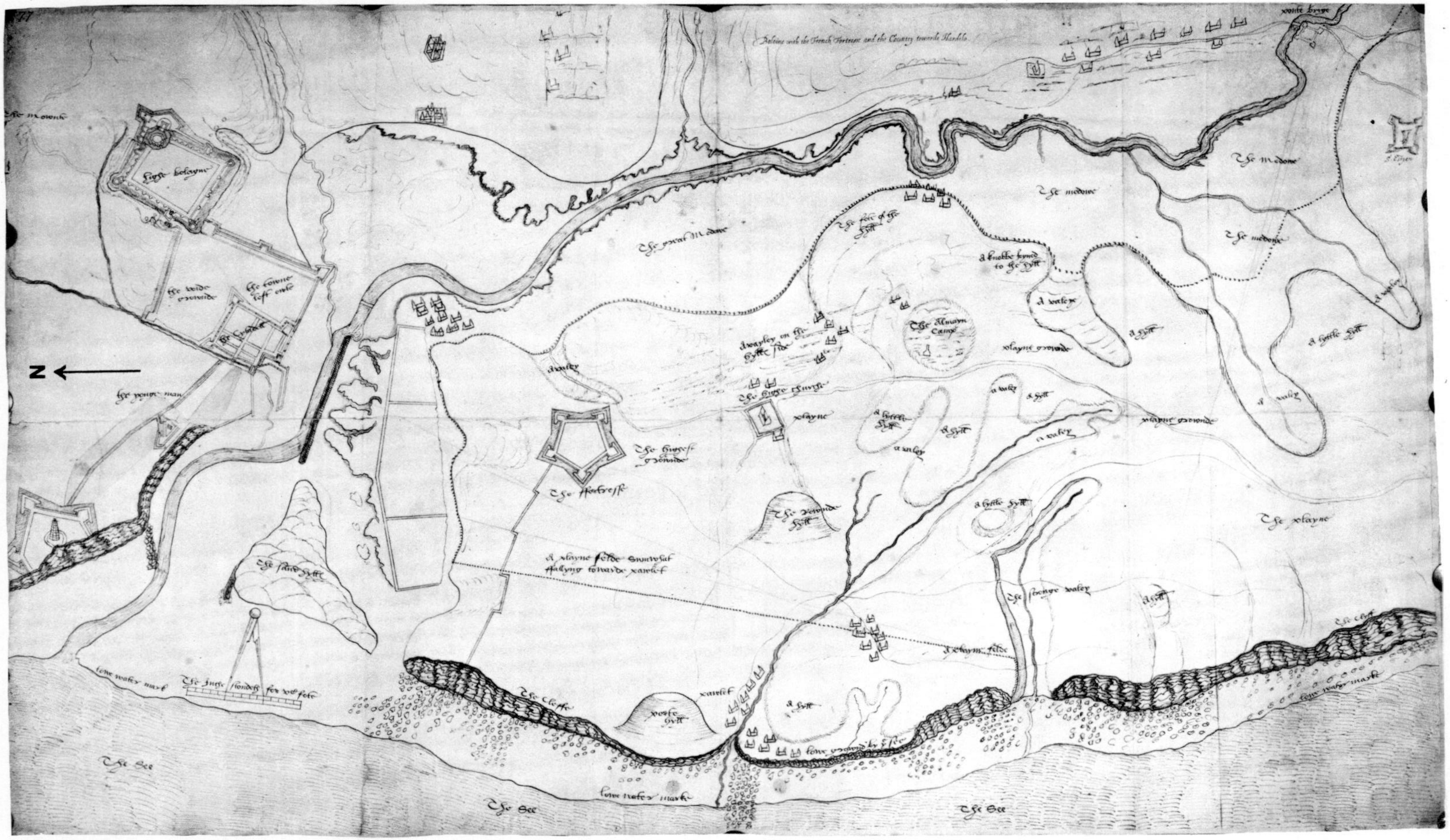

20. 'SURVEY II OF BOULOGNE'

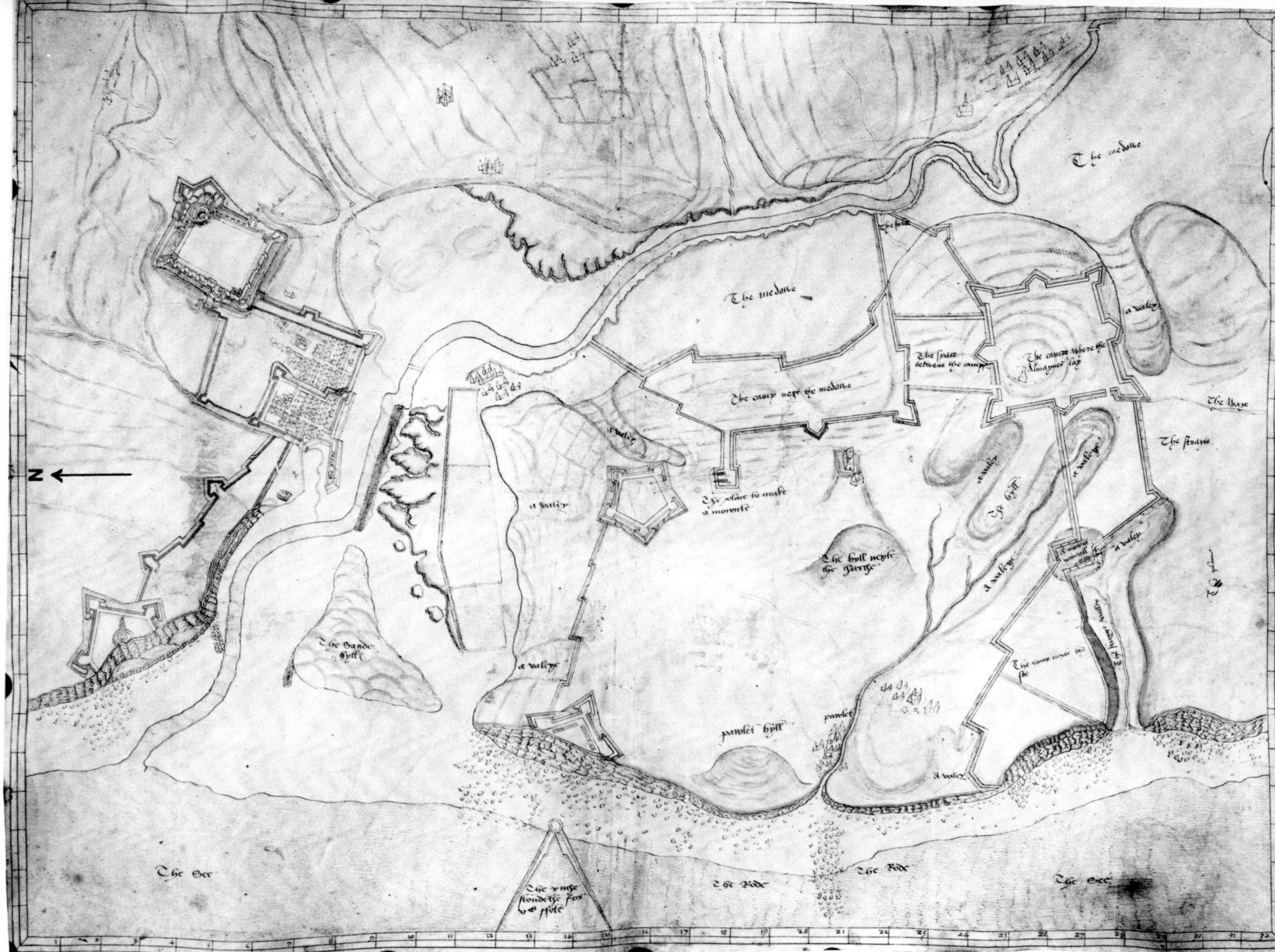

21. 'SURVEY III OF BOULOGNE'

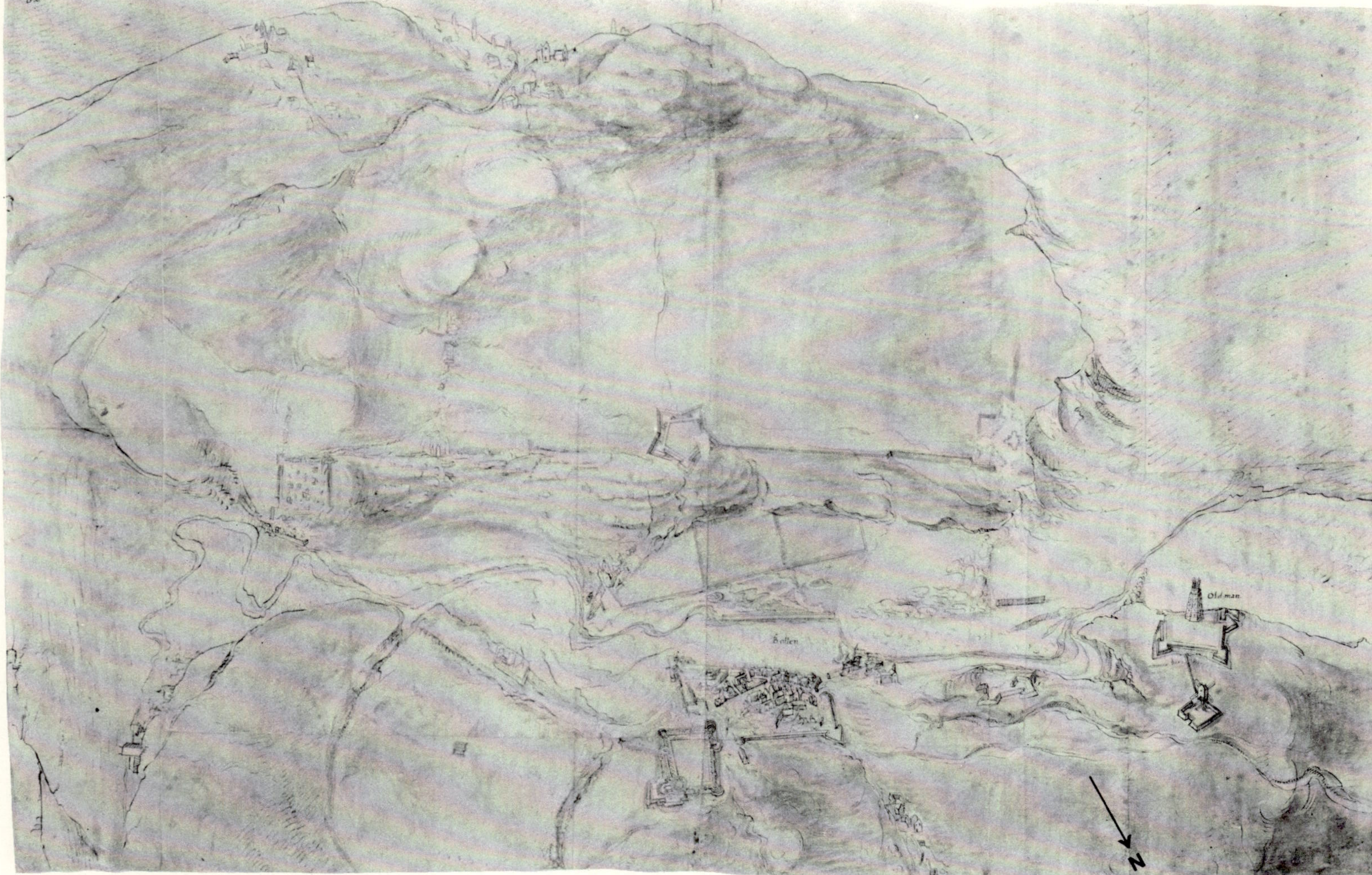

22. 'SURVEY IV OF BOULOGNE'

23. MAP OF THE PAS DE CALAIS BY NICHOLAS DE NICOLAY, 1558

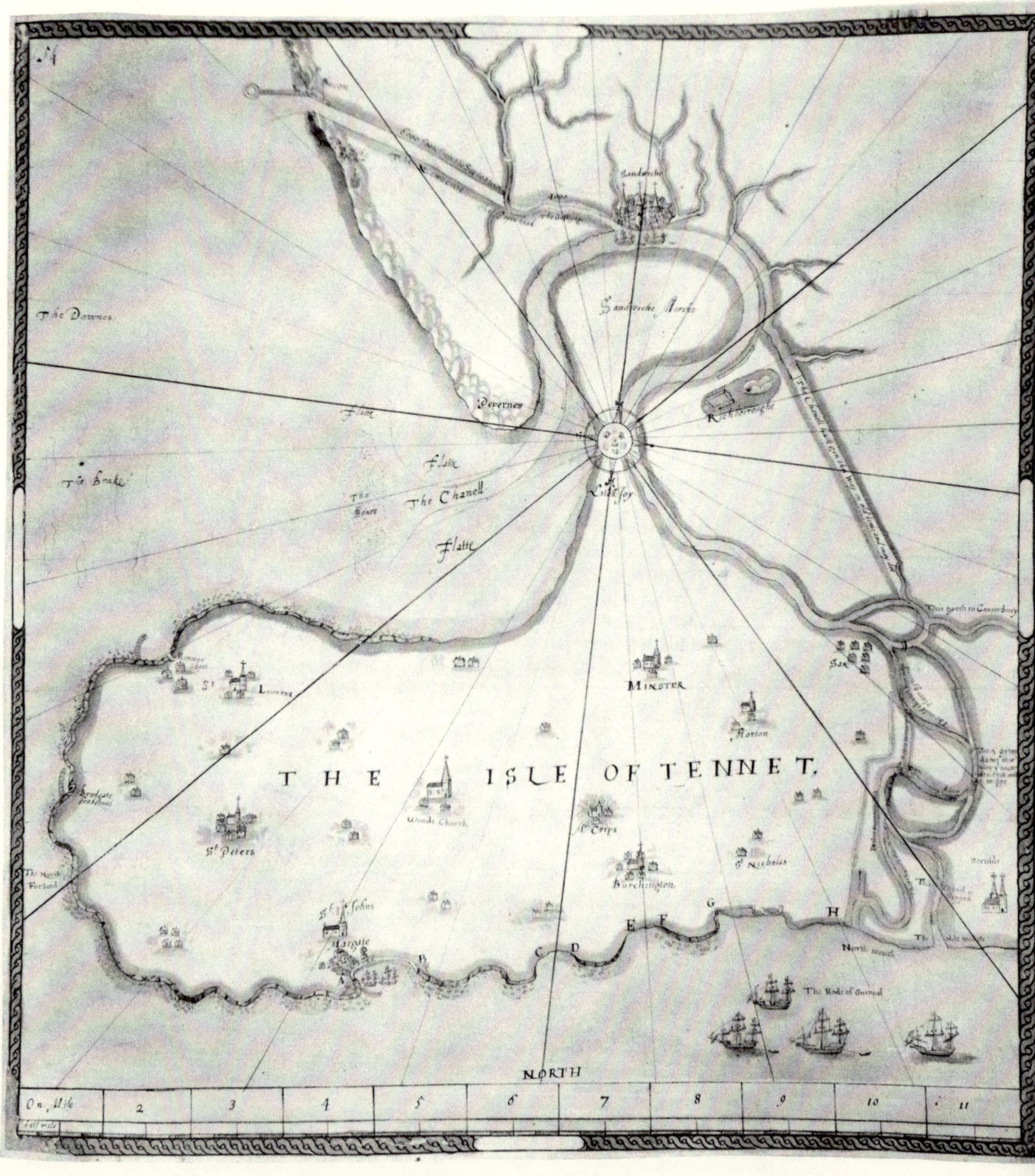

24. AN ELIZABETHAN PLAT OF THE ISLE OF THANET

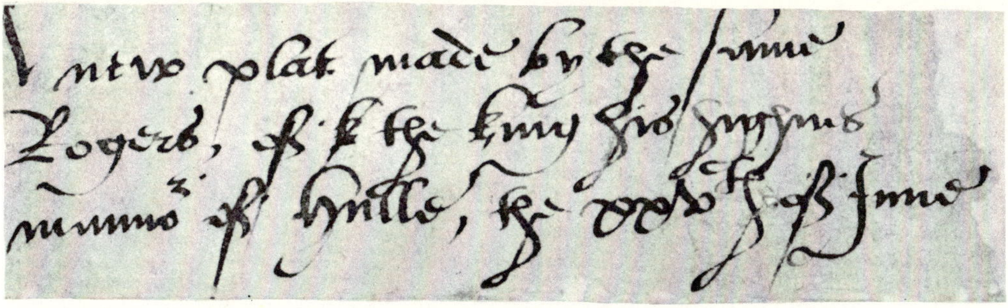

25*a*. ENDORSEMENT OF 'PLAT C OF HULL MANOR' (PLATE 12)

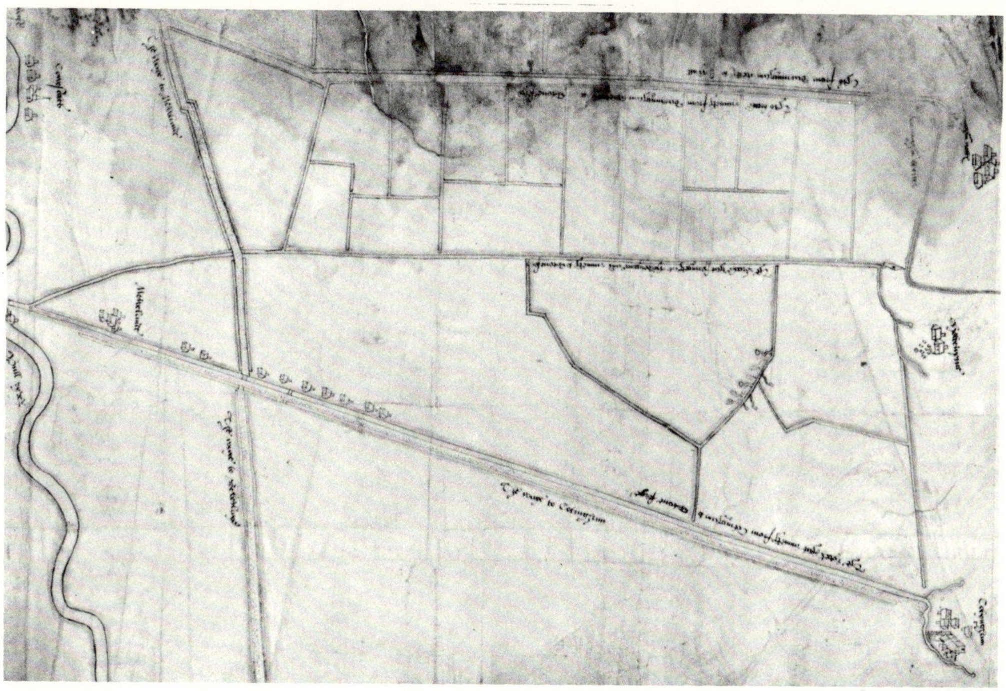

25*b*. INSCRIPTIONS FROM 'SURVEY OF HULL AREA' (PLATE 9)

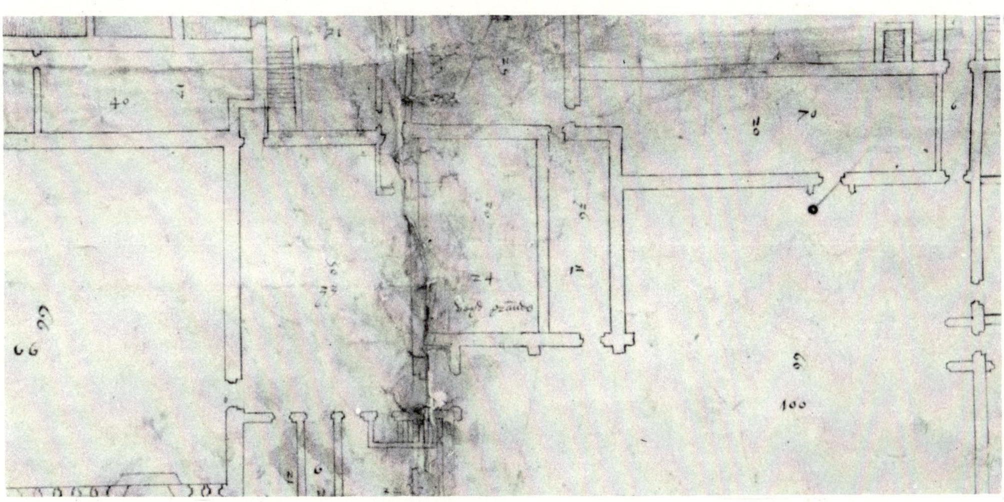

25*c*. INSCRIPTION FROM 'PLAT A OF HULL MANOR' (PLATE 10)

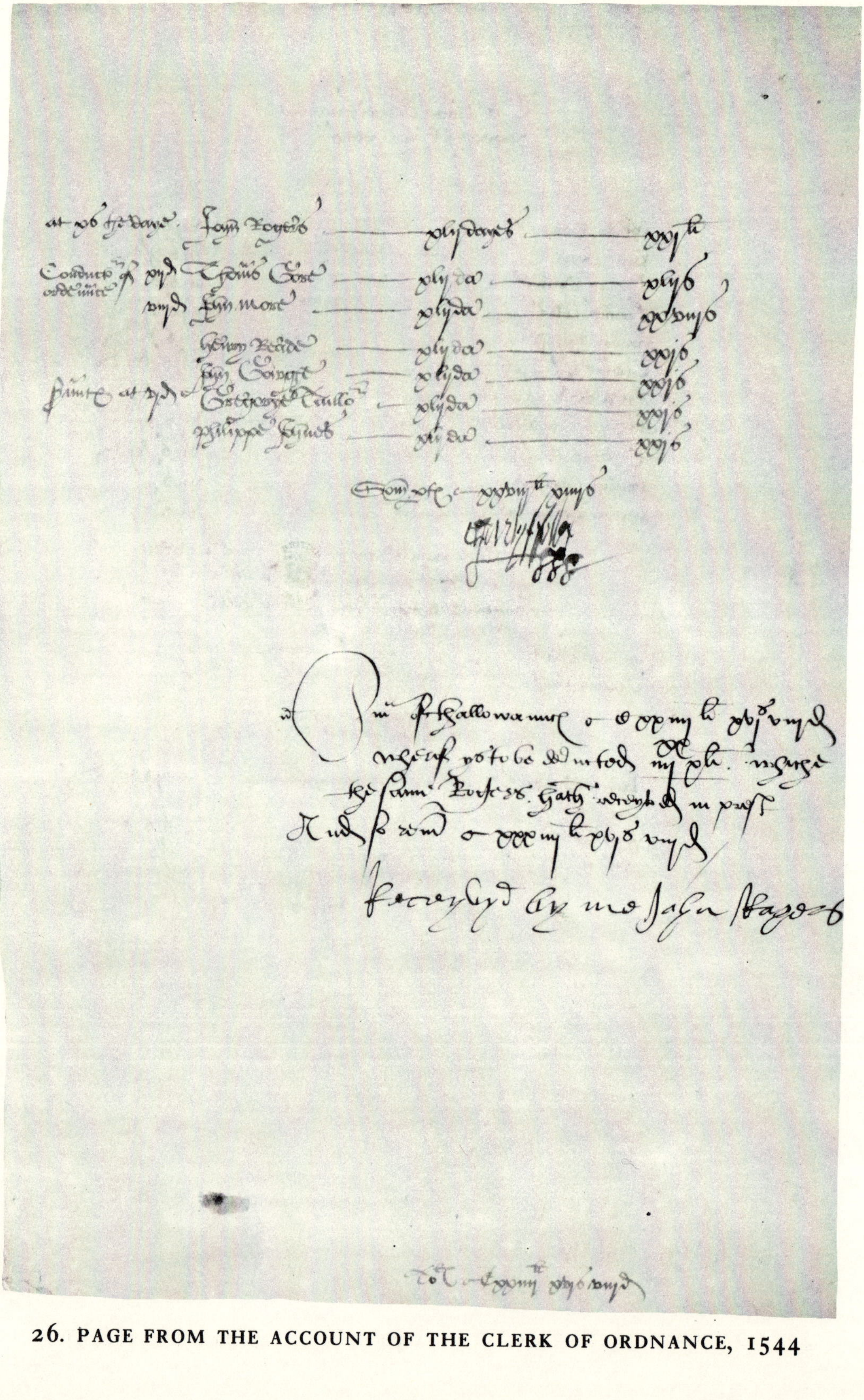

26. PAGE FROM THE ACCOUNT OF THE CLERK OF ORDNANCE, 1544

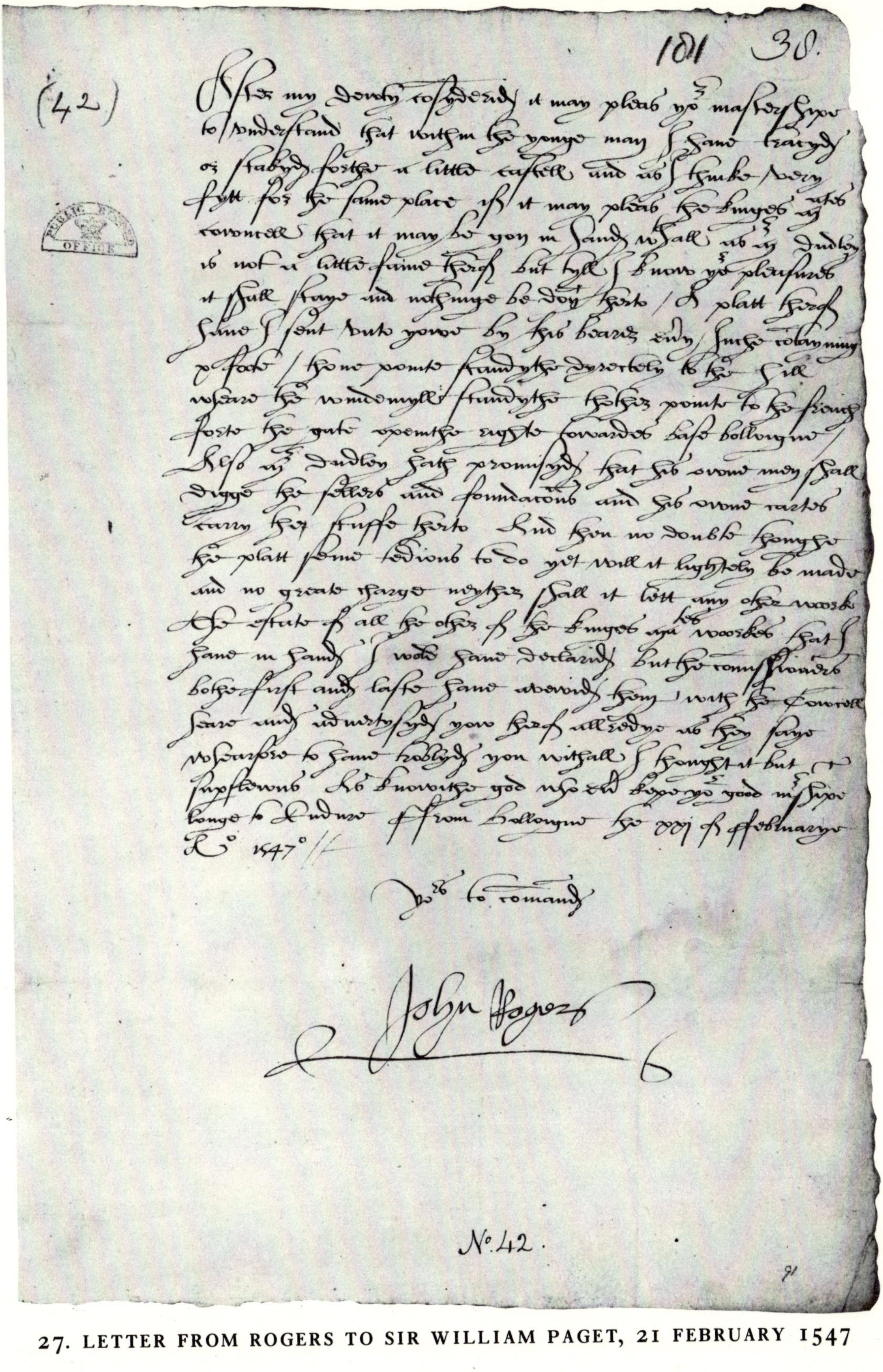

(42) 101 38.

After my dewty consyderyd it may pleas yor mastership
to understand that within the yonge may I have caryd
or stabyd forthe a little castell and as I thinke very
fytt for the same place yf it may pleas the kinges matie and
councell that it may be yn hand withall as Mr Dudley
is not a litle fayne therof but tyll I know yor pleasure
it shall staye and nothinge be don therto / A platt therof
have I sent unto yowe by this bearer and suche remayninge
perfect / the one pointe standythe directely to the hill
where the wyndemyll standythe the other pointe to the frenche
forte the gate openythe right towardes base bolloyne /
Also Mr Dudley hathe promysed that his owne men shall
digge the sellers and foundacons and his owne cartes
carry the stuffe therto And then no doubte thoughe
the platt seme tedious to do yet will it lightely be made
and no greate charge nether shall it lett any other worke
The estate of all the other of the kinges maties workes that I
have in hand I wold have declared but the commissioners
bothe first and laste have viewed them / with the Councell
here and advertysed yow therof allredye as they saye
wherefore to have troubled you withall I thought it but
superfluous As knowithe god who ever kepe yor good mastership
longe to endure / ffrom Bolloyne the xxj of ffebruarye
Ao 1547o //

Yors to comaunde

John Rogers

No. 42

27. LETTER FROM ROGERS TO SIR WILLIAM PAGET, 21 FEBRUARY 1547

408

gyve / but I thaught that yt wer a great price and he
sayed that for yt they sholde serve and delyver at Bolloyne
as many as the kynge mate wolde have / Then I prayed hym
to sende the colis as was apoynted and as for the price I
wolde not meddle / that I referred to yor grace and to the
rest of the kynge mates moast honorable councell / Also
the frenche men wolde fayne by colis at bolloyne In so
moche that at the ffyrste worde they offered xxs sterlynge
ffor a chalder so that I parseave the kynge mate may be a
great gayner yf a staple of colis wer hear / As tochynge
the reparacyons of howses ffor vitualls whiche ar styll
a doynge and never at an ende / and now yt they
required me to make ij ostes to drye malte on ffor
he hathe braught to bolloyne muche barley whearonto
belongethe fflowers to make it on / so that yf the one be
made they wyll call for the other / and so shall the charges
of the workes apere greater thearbye / as it hathe don
synce bulloyne hathe byn Englysshe hytherto / wherin
my desyer is to knowe yor grace pleasure whether
I shall doe it styll or not / all other the kynge mate affares
hear as well abowte bolloyne / as at newe haven / goethe
reasonably well fforwarde / as knowethe god who ever
preserve your grace longe to endewer / wrytten at
bolloyn the xxvij of marche /

your grace allwayes
to comande

John Rogers

28. SECOND PAGE OF LETTER, FROM ROGERS TO THE LORD PROTECTOR, 27 MARCH 15

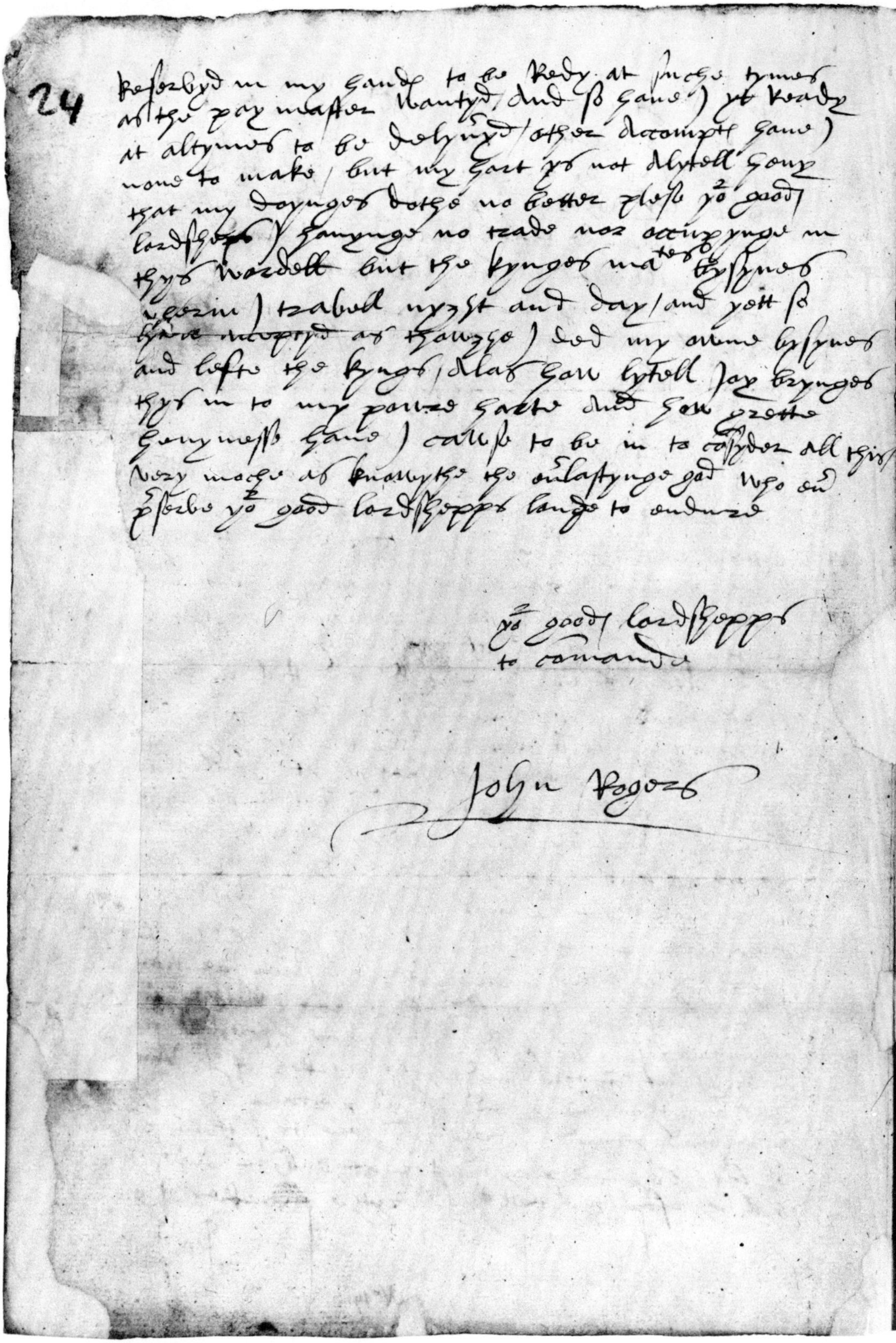
24
Reservyd in my hande to be Redy at suche tymes
as the paymaster wantyd, and so have I yet ready
at alltymes to be delyvyd/ other accompte have I
none to make, but my hart ys not a lytell sory
that my doynges dothe no better plese yor good
lordshepps I havynge no trade nor occupynge in
thys wordell but the kynges ma^tes bysynes
wherin I travell nyght and day/ and yett so
have occupyd as thowgh I ded my owne bysynes
and lefte the kyngs/ alas how lytell Joy bryngs
thys in to my powre harte and how grette
hevynesse have I cause to be in to consyder all thys
very moche as knowythe the everlastynge god who ev
preserve yor good lordshepps longe to endure

Yo^r good lordshepps
to comande

John Rogers

29. SECOND PAGE OF LETTER FROM ROGERS TO THE PRIVY COUNCIL, MAY, 1548